TRADITIONAL FISHING BOATS OF EUROPE

MIKE SMYLIE

AMBERLEY

To My Children Christoffer, Ana & Otis

First published 2013

Amberley Publishing
The Hill, Stroud
Gloucestershire, GL5 4EP

www.amberley-books.com

British Library Cataloguing in Publication Data.
A catalogue record for this book is available from the British Library.

ISBN 978 1 4456 0253 0
E-book ISBN 978 1 4456 1434 2

Typeset in 10pt on 12pt Sabon.
Typesetting and Origination by Amberley Publishing.
Printed in the UK.

Contents

Introduction

I found Theodoros at work under the shade of a big, leafy tree. He was measuring two pieces of 5 by 5 cm softwood set end to end, over another that was obviously the template. I watched for a few seconds before saying *kalimera*, trying to work out what he was doing. He nodded a reply before fetching a short length of mahogany, 15 cm square by about 30 cm. I told him Taxiarchis had sent me and he obviously recognised the name for he smiled and said something I didn't understand. He got the general idea, though, that I was staying with his nephew just up the hill. The piece of mahogany was added to the pattern he was working out. A long thin pattern with a slab of mahogany at one end. I hadn't the slightest idea what this was to be. I still don't!

I asked to photograph him at work and the slow downward movement of the head to one side, eyes slightly closed, was the Greek affirmative. He continued measuring the pieces of wood. His swept-back, greying hair matched his black and white checked shirt and his spectacles swung below his chin. A white moustache decorated his upper lip and reminded me of Ratty in *The Wind in the Willows*! A chicken wandered carelessly about the workshop – if one can describe a sheltered area under the trees as such. The nearby shed, all brick and wood with a tiled roof, was ramshackle to say the least. Inside it was stuffed with lumps of timber of all shapes and sizes and, in the middle, a huge bandsaw driven by a wide belt on an equally massive wheel, itself driven by a motor hidden away. An electric cable sneaked out of the door towards the boats sitting between beach and grass, on that area that is almost like no-man's land, in that it is neither beach nor land. The sound of an electric drill broke the silence of the late morning.

I wandered about this no-man's land, almost as careless as the chicken. The smells of the heat, the sea and a bit of cooking were delicious. A couple of old engines lie rusting among the wild flowers. Grass grows up around some of the boats, which probably haven't felt the thrill of water under their keels for many years. Others were being prepared to be rolled back down into the glittering sea only feet away. These boats were mostly *trechandiri* and *varca*, the first being the traditional Greek double-ender and the other the smaller, transom-sterned open boat. One, primed in bright orange with hull fastenings stopped, thus giving that familiar polka-dot pattern, was being cleaned out by vacuum cleaner. Paint-splattered pebbles decorate the beach.

I found Taxiarchis' *varca*, photos of which he had showed me the night before as we talked over several glasses of 'tsipouro', the strong liquor made from the skins of the *retsina* grape. This boat, like so many, had been built here on the beach. Theodoros, like his father and grandfather, had been building since he was able to remember. His last boat, sitting alongside the shed, hasn't been completed yet, and has remained untouched for seven years. Although framed up and ready for planking, work ceased on the 10-metre boat when his brother, also a boatbuilder, died. After that Theodoros had no-one to help him. Now the *trechandiri* waits for a future planking-up while the grass grows in her bilges. Maybe.

Today this small boatyard, atop the beach below the cemetery, lost in the trees but not a mile from the centre of Skiathos Town, survives through the odd repair job. Theodoros is 69 now and no-one seems keen to take over from him. Greece has joined the Great European Club, where fibreglass boats are more popular. This family used to build boats up to 33 metres in length and the islanders sailed them all over the world. For in Skiathos, situated upon trading routes since the Mycenaean civilisation during the Bronze Age, the sea has always played a major part in people's lives. Just as the Welsh captains roamed the oceans, so did the Greeks. Those islanders not keen to sail deep-sea would remain at home, looking after the older members of the family, tending the olive trees, making the *retsina* in the autumn, gleaning what is possible from a sun-drenched land. And, above all, sailing out daily in their *trechandiri* to fish. The sea was the islanders' lifeline. It was also their prison.

The waft of new paint drifts over and I awake from my thoughts. Theodoros is measuring up a rudder on a small *varca* which its owner is painting white. The sun reflects off its shiny new surface. Behind, though, an ugly fibreglass boat almost smiles. Just along the beach two young children throw stones into the water, their mother watching. By the time they learn to understand, this boatyard will have disappeared, Theodoros gone. The fibreglass boat might smile but that smile won't last long. But by then it will be too late and the children won't smile either.

2012

One of the fabulous things about European working craft, apart from the rich diversity throughout the continent, is that they tend not to obey political boundaries. This is especially so for fishing craft that work in the ocean that belongs to no national state, a common resource outside territorial limits, where conditions partly dictate the preferred vessel for the job. In other words, there's a bit of a sense of anarchism about them. Given that fishing communities were once frowned upon as 'unsuitable' to live on much sought-after beach positions, it's not at all surprising.

Four boats, from left to right: 4.6 m *Lodz Rybackie* CHA32, 5.36 m Pomeranian boat KAR4 from 1950; 8.52 m Pomeranka boat DEB5; 4.95 m Pomeranian boat CHA11 from 1956.

Inside view of DEB5, showing construction details.

The very nature of this work, and having to face all that the ocean can throw at it while at work, is, of course, the prime factor taken into account when that vessel is built, although the local conditions at home (harbour, drying estuary or beach), individual experiences and foresight as well as the oral traditions passed down through generations all play a part in the final design. Thus almost everywhere we see an overlap at country borders, such as the typical Scandinavian boats that occur throughout the western and northern Baltic, the Pomeranian beach boats of the former East Germany and Poland, the clumsy-looking beach boats of Holland, Belgium and France, the Basque country craft, the Spanish/French *barca catalan/barque catalane* and the French/Italian *gozzi*, to name a few examples.

Europe's boatbuilding tradition has evolved from two distinct styles – the shell-first carvel vessels of the Eastern Mediterranean introduced further west by the Veneti and the shell-first clinker craft of Scandinavian and, before them, Saxon traditions. Added to the general picture are the Celtic skin boats, the traditions of the Sami and peculiar craft such as the Sardinian *fassone*, the reed boat of the western lagoons.

In other lagoons various flat-bottomed craft have developed from early log boats, some with longitudinal planking in the bottom and others, more rarely, transverse. From the Baltic canoes, the flatners of Somerset, the *bettes* of Southern France, the sandolo of Southern Italy, the sandula of Croatia down to the Messolongi canoes and *kourita* of Greece, there is a common heritage, appearing almost as luck, though in reality it's down to the similar conditions of work.

There are some 204,000 miles of coastline in Europe, a distance only exceeded by that of North America, where a huge portion is in the Arctic region, frozen for much of the year. Ice only generally affected the Baltic and inland waters. Although fifteen years of research has gone into this book it hasn't, due to time and expense, been possible to travel the entire coast. Some countries have been left out for obvious reasons – Russia, Kalingrad, Bosnia, Montenegro, Albania, the Black Sea and Turkey. Out of these, Albania should have been included, though the excuse is that although the country had a healthy, though small, fishing fleet, these were craft similar to those to the north and south. Britain and Ireland have not been included as my earlier book *Traditional Fishing Boats of Britain & Ireland* has recently been reprinted to compliment this book. Neither is this book a complete and concise list of vessels, as there are just too many. I'm happy to say, though, that most have been included.

However, here there is always a problem! Academics insist that boat types are recognised in categories whereas boatbuilders and fishermen didn't quite see it that way. Thus with a tweak here and there one boat type might be slightly different to another, though both are regarded as being the same type. Some might prefer a different rig to others. This always has to be taken into account where, often, no two boats are the same even when built by the same hand. To most fishermen they were simply 'the boats'. Labels are our own invention.

Over the last 20 years there has been an abundance of replicas and restorations in almost all European countries as the awareness of the importance of maritime heritage has grown. Sadly, at the same time, the bureaucrats in Brussels have been

The *lodz Rybackie* – literally 'fishermen's boat' – from the Vistula Lagoon.

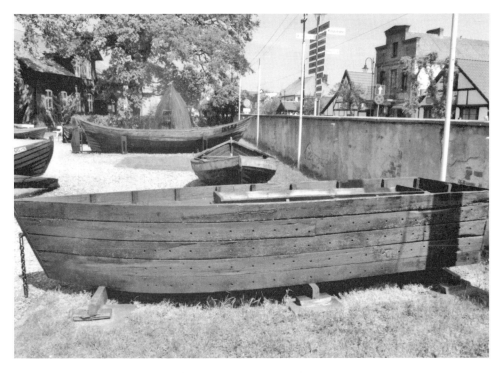

A fish storage vessel for holding fish, also from the Vistula Lagoon.

decommissioning fishing vessels and enforcing a scrapping policy. Thus hundreds of historically important, seaworthy and beautiful fishing craft have unjustly disappeared, pulled apart in harbours, burned on beaches or left to slowly rot and die.

Museums in almost all the countries mentioned have exhibitions on fishing-related subjects while some countries have museums dedicated only to fishing. Some of these have proved live-saving at times during the construction of this book.

Dozens of people have contributed in one form or another to this book, as would be expected, as the subject is pretty vast. This has come through a mixture of information, photographs, reminiscences or simply encouragement, for which I will eternally be grateful to these good people and establishments. These, in a totally random order, are: Soren Byskov and the Fisheri-og-Sofartsmuseet, Esbjerg, Denmark; National Visserijmuseum of Oostduinkerke, Belgium; Daniel Bosser; the Dutch National Fisheries Museum, Vlaardingen; Cas & Kirsten Trouwborst; Hermann Ostermann; Reiner Schlimme; Hans-Christian Rieck; Vicco Meyer; Dragana Lucija Ratkovic of the Rovinj Eco Museum (see www.batana.org); Hannu Vartiainen and the Rauma Maritime Museum, Finland; Pablo Carrera and the Museo do Mar de Galicia, Vigo, Spain; the Bermeo Fishing Museum, Bermeo, Spain; Marga-Leena Hanninen and the Finnish National Board of Antiquities; Tanel Laan and the Emajoji River Barge Association, Estonia; Jaak Tambets; Elo Raspel; Giovanni Panella; the late John Bushell; Michael Craine; Peter Radclyffe; Stewart Hyder; the Maritime Museum of Cesenatico; the Barcelona Maritime Museum; Joseph Muscat, curator of the Maritime Museum of Malta; Petros Kounouklas; the Scottish Fisheries Museum, Anstruther; Romas Adomavicius, historian of the Lithuanian Sea Museum; Adrian Osler; Kostas Damianidis; Velimir Salamon; Robert Prescott; John Robinson; Thedo Fruithof; Robert Simper; Bart Wordsworth; Bjorn Lingener of the Hardanger Fartoyvernsenter, Norway; Aleksander Celarek; Jan Pentreath; Luke Powell; Marina Sundmacher and Sandro Russo, both of whom I 'met' through the photo site Panoramio; Kundi, Chano and Ildefonoso from El Cotillo and Pepiyo from Corralejo, all of Fuerteventura; and Campbell McCutcheon, Chris Skal, Louis Archard and the team at Amberley.

PART ONE

SCANDINAVIA AND THE BALTIC SEA

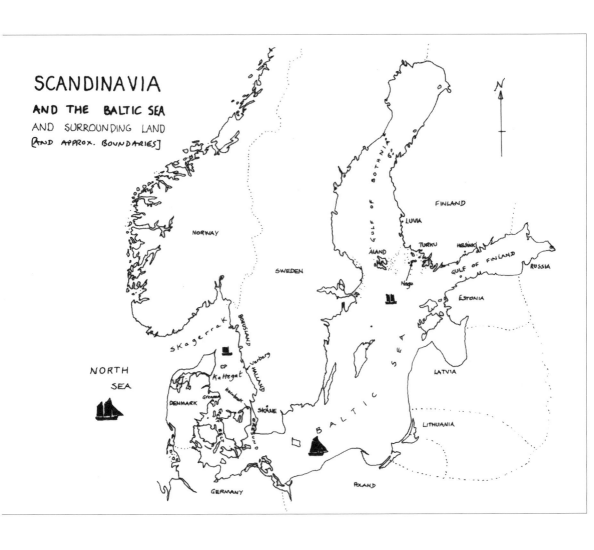

SCANDINAVIA

AND THE BALTIC SEA
AND SURROUNDING LAND
[AND APPROX. BOUNDARIES]

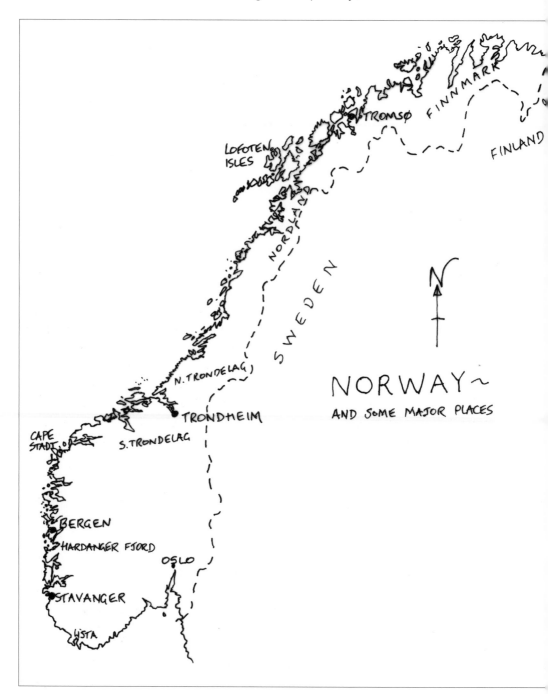

CHAPTER 1

Norway

The North Cape to the Skagerrak

The Norwegian coast is long, some 15,700 miles long if you take the islands into account, and is the largest coastline of any European country, the second largest in the world according to some reports. Though the mainland only accounts for some ten per cent of this length, its coast is characterised by high, snow-capped mountains that plunge down into the deep fjords penetrating many miles into the coast. Offshore, it is the 75,000 or more islands that add up to create such a vast coastline. These islands act as a chain of protection to much of the coast and form many sheltered bays on the rocky islands where boats could be beached, as well as protecting inland waterway routes. Here, too, the climate is warmed by the Gulf Stream, which also makes it rich in fish. Onshore, although coastal settlements have grown on the shores of many of the fjords, higher up the land is generally unfertile except in the less mountainous south, where most of the country's agriculture is centred. Some say that it was this lack of good growing land that led the Viking sailors to set sail westwards in their *knarrs*, to find land where they could grow their food and tend their animals, as well as continue fishing. Britain must have been an ideal discovery with its lush fields, forests thick in timber and rich fishing grounds just offshore. Whether this was merely a stepping-off point for further explorations still remains a mystery, though Vikings did sail to the Faroes, Iceland and Greenland, so it takes little imagination to believe they did arrive in Newfoundland long before John Cabot or any other European pioneers.

What the Vikings did have, though, were abundant supplies of good timber growing slowly on the hillsides, timber perfect for boatbuilding such as pine and spruce. That the Vikings developed their skills in boatbuilding, unmatched by others in Europe at that time, is undisputed. Along with this timber, they had skills drawn from generations before them. Their craft were strongly built in a clinker tradition, yet cleverly designed so that parts that wore out – such as the keel – could be easily replaced. They were almost all double-ended, with the characteristic high bows and squaresails, and many of these features continued right down to present times. The earliest known example of such a boat, though representing pre-Viking boatbuilding, is the Hjortspring boat, which was discovered on the Danish island of Als and dates from the fourth century BC.

According to Bernhard Faeroyvik, who measured a host of traditional boats which were fast disappearing in the period between 1924 and 1950, there were seven distinct types of vessels working the Norwegian coast up to the 1920s, by which time motorised vessels had made many of the traditional types of fishing craft obsolete. Generally, he represented these groups as: Coastal Sami craft, Nordlands boats, Trøndelag craft, NW coast types, mid-west coast types, SW coast types and SE coast types. Although there are many similarities between the types, there does to seem to be a split between the boats of the southeast – the length of coast between the southern tip of the country and the Swedish border – where heavy oak-built, sprit-rigged boats were common, and throughout the rest of the coast to the north where lighter craft, built of pine or spruce, were often rowed instead of being sailed. We shall look at each type in turn. However, although the majority of Norwegian craft come in various shapes and sizes, it must be remembered that they are not only fishing vessels, but craft used for all manner of day-to-day work such as carrying animals and goods as well as humans. Most of these craft were originally referred to by the number of oars they carried or the number of 'rooms' between the frames (the place where the oarsman sits) in the Scandinavian custom. And throughout Scandinavia, as will be seen, such craft were also adapted as church boats, being specifically to collect and transport people to their weekly ritual of church-going. These craft have, in some cases, been immensely long – over 120 feet – to carry a congregation in excess of 100 persons, propelled by 30 pairs of oars.

NORWEGIAN FISHING CRAFT

THE SAMI BOATS

The Sami live throughout much of northern Norway, especially Finnmark, as well as into northern Finland and parts of northern Russia, in an area collectively known as Lapland. The boatbuilding traditions of the Coastal Sami are thought to stretch back as far as the Late Iron Age and their boats were built of sewn-planking instead of using iron spikes, using local timber such as spruce. Though bast rope was used to sew the planks together – it has been suggested that boats consisted of no more than four wide planks – it is thought that the early Sami used roots, hemp ropes or reindeer sinews for their early boats and moss to caulk them. According to the Icelander Snorri Sturluson, two 24-oared sewn ships were built by the Sami for King Sigurd Slembe in the twelfth century, which suggests that sewn boats were still being built when the Vikings set sail to Britain in the late eighth century. Sewn boats continued in use among the rivers of Finnmark and Lapland, where there was quite a diversity of small river craft such as canoes, dugouts and other such craft.

Boatbuilding traditions throughout the world have evolved from the early roots – the logboat, raft, bark boat and skin boat. The sewn-plank boat in many parts developed from the simple dugout log, in that planks were added to this. The Sami

A typical Sami boat with three planks either side. This, usually with a small squaresail, was the workhorse of the north, fishing, transporting and generally used for everything by the ocean-based communities.

tradition is generally believed to have been slightly different, in that the planks were sewn directly onto the bottom or keel of the boat. This is a shell-first construction, in that the hull is planked first before framing is added internally to strengthen the vessel. Such is the way that all Norwegian craft were built until the twentieth century.

An early Sami fishing boat of the extreme north was the *schnjaka*, which was used to net-fish for cod and herring. It has been suggested that this is in fact a Russian vessel whose name is derived from the Norwegian *snekke*. It was a sewn-planked vessel of some 12 metres, rigged with a single squaresail. The holes that the twine runs through – a double row – were pegged and some iron nails were used in the hood ends, according to one example excavated. An example of a smaller six-oared boat from the late nineteenth century still exists in the Tromsø Museum. It is believed that vessels such as these have been in use for well over 1,500 years. However, in that century, vessels were brought in from Nordlands which might have led Faeroyvik to believe that Sami craft were influenced by the general Norwegian boatbuilding tradition. Others say the exact reverse is true and that the Coastal Sami craft influenced the boatbuilders to the south.

NORDLANDS BOAT

One of the most spectacular of all the Norwegian craft comes from the Nordlands – a stretch of the coast mostly within the Arctic Circle which has been described, among other things, as 'the land of high flames', a reference to the long summer days and the Northern Lights in the winter. The vessel is the *Nordlandsbåt* – literally a 'Nordlands boat' – a vessel used for transport and fishing in these exposed waters and said to be the last boat in Europe working under a squaresail. These craft are normally referred to by the number of 'rooms' and thus they came in various sizes. The smallest is the *tororing*, a two-man vessel, while the *treroring* is a three-man boat. At about 15 feet in length, both of these were fishing boats using a hand-line close to home. At the other end of the scale were the 45-foot *fembøring*, the six- or seven-man craft used for the deep-sea cod fishery. In between was the five-man *attring*, used for the long-line fishery.

The oldest *Nordlandsbåt* still surviving is about 200 years old, though many others still exist, such as 'Opreisningen', a 44-foot *fembøring* from the very early twentieth century which is now exhibited in Oslo. Furthermore, many new vessels have been built over the last 30 years, many based on patterns from the late nineteenth century. With their high bows, low freeboard amidships and *lofting* – the removable stern accommodation for four fishermen – the *fembøring* have participated in the Lofoten cod fishery since Viking times. Here, spawning cod swim in huge shoals into the narrow

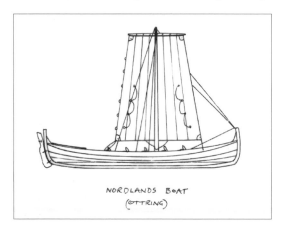

NORDLANDS BOAT
(OTTRING)

sounds of the Lofoten Islands in late March, for which fishermen come from all over Norway and which accounted for some 70 per cent of Norway's fish export. They would remove the *lofting* and carry it ashore to camp, a method rarely seen anywhere else in Europe, where camper fishermen generally built their temporary lodgings from old material, such as the Scottish west coasters did.

An open five-man Nordlands boat – or *attring* – used for long-lining in what must be icy waters, judging by the snowy peaks in the background.

Nordlandsbåts are built wholly from pine and covered in brown pine tar to protect them. Many of the larger boats are rigged with a squaresail and a topsail (*storseil* and *toppseil*), nowadays both made from linen, though woven wool sails were once common. The smaller boats would only have the one squaresail. Rigging was made from horsehair or hemp. They were renowned for their seaworthiness, which some attribute to the shape of the hull between bilge and keel which sucks a spiral of air beneath the boat when heeled at speed, thus giving the boat lift. Obviously running before the wind and reaching is simple for these craft, yet they do also have a good windward performance. It is said that the fishermen used a loose ballast of huge round stones in the bilges which, in the event of a capsize in heavy weather, simply rolled out into the sea, thus enabling the wooden vessel to stay afloat. However, with sea temperatures down to zero, any fisherman wouldn't last long once immersed in these waters.

A word about the Lofoten cod fishery. This really was the greatest cod fishery in the world and attracted fishermen from all over. It's also a very old fishery, illustrated by the fact that in about 1120 King Oystein built *rorbu* cabins around the coast for visiting fishermen, which in turn brought in even more fishermen. The fishery has been the subject of legislation over the years, with long-lines banned in 1644 in favour of hand-lines. When gill-nets were introduced they, too, were banned for a brief period. In 1786 both were legalised. Then, in 1816, the allowed fishing time was regulated and visiting fishermen were made to rent *rorbu* cabins in advance, deciding on the type of fishing they wished to follow, and thus stick to specific areas. This became very unpopular as the rigid rules were implemented by inspectors who were also the landowners, thus taking advantage of them. The ruling also took no account of the random movement of the cod, which did not distribute themselves equally around the various villages. Some fishers did very well while others found no fish in their area, with obvious results. However, these regulations were finally lifted in 1857 and all the seas were accessible to all, though the state retained the rights of inspection. But it

Larger Nordland boats clearly showing their *lofting*, the removable shelter. These are larger seven-man – or *femboring* – boats with very square sails, which were made from wool until the nineteenth century.

Another *femtøring* near the old Finnish town of Petsamo before they lost access to the northern coast. Unusually, this is rigged with a gaff main and small gaff mizzen.

wasn't until the 1930s that the fishermen were able to obtain a minimum price under the Raw Fish Act of 1938.

Once caught, the fish were dried. Much of the catch was bought by the captains of *jakts*, large trading vessels built in several places along the coast (Hardanger and Sunnfjord were renowned for their *jakts*), which arrived with goods such as salted meat, grain and butter from the south. This they bartered for cod, which was loaded aboard and salted in the hold. They would subsequently hire suitable cliffs from a local farmer upon which to dry the fish in the sun before it was exported to Portugal, the main customer. The same *jakts* would carry exported herring to the Baltic at other times of the year.

There is a photograph in the archives of the Finnish National Board of Antiquities in Helsinki showing one such *Nordlandsboat* rigged as a gaff ketch with jib and foresail set on a bowsprit. The mainsail appears to be loose-footed while the small mizzen is set over the *lofting*. Whether this rig was normal is uncertain, though it is said to have been introduced after English sailing trawlers were bought by Norwegian fishermen from further south. The vessel is photographed at Petsamo, a one-time Finnish fishing station on the Norwegian Sea coast, and there is a doubt as to which country this vessel belonged to. Indeed, judging by the lettering on the bow, it is possible it was Russian registered.

BOATS OF THE TRØNDELAG REGION

The boats of the Trøndelag region to the south of Nordlands were indeed similar to those from the Nordlands. The biggest of these was called a *femtøring* or *storebåt* (big boat), with the largest *femtøring* being used to sail north to participate in the Lofoten cod fishery, these boats also being referred to as *Lofotbåt* for obvious reasons. Smaller eight-oared boats were used for the cod and herring fishing in the local vicinity. One of the largest of the *femtøring* was measured by Bernhard Faeroyvik, a boat that had been built in 1892 by Hans Severin Berdal from Afjorden, which was recognised as the principal boatbuilding centre of Trondelag. This vessel, at some 40 feet in length, would quite likely have sailed up to Lofoten, which was 200 miles away. It was rigged with a large squaresail with a smaller topsail. Some later adopted the gaff rig.

NORTH-WEST TYPES

The north-west boat types came from the part of the coast between Cape Stadt in the south and the Trondheimsfjord, where the fishermen were said to have to voyage further away from the coast to reach the best fishing grounds in their quest for cod and ling. However, within this part of the coast, there are two areas – Sunnmore and Nordmore – which have their own types, though Faeroyvik classes them together as he regarded them as having enough common traits to do so.

The Nordmore boats had to cope with treacherous, shallow waters with off-lying breakers over skerries and rocks, and here, it seems, there is an influence from the southern boats as well as those from the north. A typical *Nordmorsbåten* had long overhangs on a relatively short but deep keel. In the north of the area they were rigged with one squaresail, while to the south they adopted the dipping lug after the boats from Sunnmore. Sizes ranged from four to twelve oars, the bigger boats being used to fish for cod. When they were fishing in the same grounds, the Trøndelag fishermen called the Nordmore boats *geitbat* or 'goat boats', a name that stuck among the Nordmore crews.

The boats of the Sunnmore fishermen had a 'cod's head, mackerel tail' shape and were much shallower and beamy than those from the north. For some reason, thought to have come about when boats increased in size, they referred to a ten-oared boat as an *ottring* or eight-oared boat and, similarly, the six-oared boat was called a *treroding* or three-oared boat. They also had a strange way of planking, in that two narrow strakes running aft from the stem are joined to one wider strake running back to the sternpost. This is said to make the boat elastic so that it is able to rise over oncoming waves. Fishermen supposedly tested the elasticity of a new boat by shaking the stem head up and down to see if it would 'wag its tail'! Such a construction, though, has the disadvantage of being weak around the point that the planks meet and, as the nets were carried forward and the catch aft, tended to hog badly and, in extreme cases, break up while at sea. The *Sunnmorsottring* was one boat measured by Faeroyvik which was constructed in this way. However, at a little over 40 feet in length, this is regarded as being small for an eight-oared vessel. Being narrow aft and without any flatness in the floors, this was generally thought to be unsafe in a following wind. The dipping lug had two hand reefs (*handsyfte*) which could be activated quickly to prevent a capsize.

Around 1870 they adopted a new shape, with strakes that reached from bow to stern. This influence came from the Lista and Hardanger boats that we will discover more about later. Furthermore, the older Sunnmore boats had, at least since the middle of the eighteenth century, been rigged with a dipping lug, until, that is, the adoption of the new shape when they also adopted a two-masted standing lug or gaff rig. One factor of the new boats that didn't impress the fishermen was that the older boats, like the *Nordlandsbåt*, carried ballast in the form of stones that rolled out if the vessel capsized so that the boat did not sink, at least enabling the fishermen to hang on. The new boats, when capsizing, became submerged and sank. However, the new shape was regarded as being much safer at sea.

MID-WEST TYPES

South of Cape Stadt we come across the mid-west types and this type covers the coast down as far as the Osterfjord, just to the north of Bergen. The bulk of the fishing took part in the north of the area, between the Sunnfjord and the Nordfjord, the boats of the latter having said to have been among the best along the coast while the boats built in Gloppen were supposedly the best of the Nordfjord. These were deemed to have a higher freeboard and a fuller hull aft, taking the shape away from the 'cod's head, mackerel tail' concept, and making them more seaworthy. All of the mid-west types had four strakes, until the boat size increased so that planks wide enough were unobtainable, thus meaning that five strakes had to be fitted. One of the largest of the four-strake boats was the *femkeiping* or 'five rowlock boat', used to catch cod and herring, though this type might just as likely be found to be carrying hay or other freight. Some were used as church boats while an even larger twelve-oared boat from Gloppen was measured by Faeroyvik at over 30 feet. The dipping lug was the favoured rig again, having been imported from further north during the nineteenth century.

Smaller four-oared boats were used for small-scale fishing. For herring fishing with a seine-net, small transom-sterned boats, of some 20–25 feet in length, were sailed to the off-lying fishing grounds before the seine was set out from a suitable shore. The transom was broad to carry the net, which often ran out over a roller, and they were originally square-sailed, though these were superseded with the gaff rig.

In the waters off the Sunnfjord there was a great herring fishery, mostly in the early part of the year into early spring, which, like the Lofoten cod fishery, attracted people – fishermen and farmers – from all around. Most of this fish was caught using drift-nets or seines, the latter being similar to those just described. The offshore boats were flattish in the bottom and able to carry plenty of weight and were particularly suited to the herring. As in the vast majority of Norwegian fishing, many only used the sail to reach the fishing grounds, after which the mast was lowered, while when fishing close inshore, the boats would often solely be rowed. The design of boats altered substantially in the latter part of the nineteenth century when fishermen went further offshore to fish long-lines, at which time they adopted the boats of the Nordfjord. Before that, when fishermen had arrived from outside the area with their modern *Lista*-type craft, the Sunnfjord men had continued to favour their own craft for the herring fishery.

Typical herring boats from the south-west of Norway in the late 1800s or early 1900s. They continued using seine-nets until the late nineteenth century. Note the rollers on the sterns of two of the outer craft.

SOUTH-WEST TYPES

From the waters around Bergen to the southern tip of the country, Faeroyvik identified the boats as belonging to the south-west type. These boats were all three-strake craft and huge numbers of four-oared boats followed the inshore fishery. These could be built in a week, whereas larger six-oared boats that worked long-lines further out to sea took at least two weeks and the eight-oared boats that sailed up to Sunnfjord for the late winter/early spring herring, as mentioned above, might take a month or more. Up to the nineteenth century these were square-rigged craft, until the sprit rig gained favour. However, the sprit was deemed unsuited to the larger narrow boats and so, about 1860, the *Lista*-type boat took precedent after the Hardanger boatbuilders copied the shape.

SOUTH-EAST TYPES

According to Faeroyvik, the *Lista* boat originated from Hardanger, when local boatbuilder Gjert Gundersen moved south to the Lista peninsula. The boats of the south-east type were sturdy, oak-built craft rigged with a spritsail and these sailed long distances, both north and southwards, in search of fish. These were narrow in shape, deep hulled and built of numerous strakes with permanent thwarts acting as cross-members, compared to the habit on the west and north coasts of fitting loose thwarts with crossbeams to strengthen the hull. Being heavy, they were good sailers, though not easy to row. However, Gundersen experimented, using his background skills of light-weight boatbuilding and combining these with influences from the south. He wanted to build a relatively light boat, but one that sailed as well as the south-east

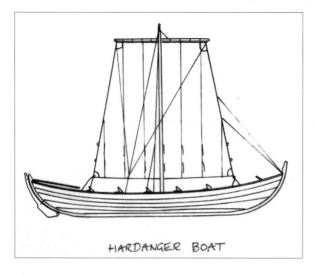

HARDANGER BOAT

types. Slowly the *Lista* boat evolved, a mixture of west and east characteristics, and these boats eventually influenced the design of boats all along the north and west coasts of the country.

Gaff rig was adopted and by 1900 the decking-over of traditional types began to appear, at first in the south but gradually reaching up along the coast. However, this is nowhere near the end of the story. Prior to this, in 1861, two Swedish smacks had begun long-line fishing off the coast of Sunnmore, employing local fishermen. With good catches, the merchants of Ålesund had begun to copy the idea and build half-decked smacks, the first being built by Carl Joachim Haasted in the autumn of that year. More soon appeared and these boats, among the first decked boats in the country, fished offshore while the smaller traditional types remained at the inshore fishing.

English smacks were bought in around 1890, after steam trawlers were adopted into the English and Scottish fleets across the North Sea. These 'Englishmen', as they were called, so impressed the fishermen that they began to build their own boats on similar lines. For the first time these were carvel-built craft, a method previously unknown in practice in Norwegian small craft construction. By the early part of the twentieth century, when some of the larger boats were being fitted with motors, fishing cutters such as the *Hardanger kutter* took over. These cutters were an improvement on the smacks, with a rounded counter-stern and a pronounced sheer. Such a vessel is the 1915-built *'Vikingen'*, built by Lars B. Hauge in Hatlestrand, Hardanger, as a fish buying boat but which worked the deep-sea fishing and, at other times, long-lines and nets along the coast from the south up as far as Finnmark. Ketch-rigged at new, she also had a 16hp Volda-Hein hot-bulb engine. To the south the hull shape was more traditional, illustrated by the double-ended *'Paddy'*, which was built in 1916 at Mandal but completely modified and lengthened in the 1950s, at which time her stern was also modelled into that of what became known as the typical Danish style. The 1916-built *Faxsen* is another cutter, built in Mosjoen with an Alpha engine, although this was replaced with the current single-cylinder Wichmann semi-diesel in 1939. The vessel was also lengthened in the 1950s. Today these three vessels sail, as do many other traditional Norwegian craft, though many of the smaller boats are replicas. However, with such an immensely important and influential tradition of boatbuilding, it's hardly surprising that there are, today, many groups, associations and individuals whose aims are to secure the place in history for these traditions that were, until Faeroyvik began his work, in danger of disappearing altogether.

A traditional Hardanger boat, said to be the finest that the Norwegian boatbuilders built. (*Photo: Stavanger Museum*)

Although said to be a Hardanger boat, it has been suggested that this, because of its curved ends, is from the Trondheim region to the north.

A Hardanger boat at the Brest Maritime Festival in 2000.

Above left: The 1916-built cutter '*Faxsen*', also sailing from Brest to Douarnenez as part of the fleet in 2008. Like many cutters, this boat was lengthened in the 1950s to reflect the changing nature of their work from fishing to coasting.

Above right: '*Faxsen*' once again surging along.

Above left: The 1915-built cutter '*Vikingen*'. (*Photo: Hardanger Ships Preservation Centre*)

Above right: The canoe-sterned cutter '*Paddy*' in 2005. (*Photo: Hardanger Ships Preservation Centre*)

Norwegian craftsmen demonstrating their skills at the Brest Maritime Festival in 2008.

SOURCES

Much of the above information comes from the best source, *Inshore Craft of Norway*, edited and translated by Arne Emil Christensen from a manuscript by Bernhard and Oystein Faeroyvik. Other sources are the Norsk Fiskevaermuseum (Norwegian Fishing Village Museum) at Lofoten and a paper entitled *Carvel Building in Norway 1800–1990* by Tom Rasmussen and Asmund Kristiansen, published by the Hardanger Fartoyvernsenter (Hardanger Ships Preservation Centre) at Norheimsund. The latter has also been helpful in the preparation of this chapter. Norway has several museums dedicated to maritime subjects, especially in Oslo, Bergen and Stavanger, and traditional boatbuilding thrives.

CHAPTER 2

Sweden

The Skagerrak to the Gulf of Bothnia

As far as fishing is concerned, Sweden is a country split – split between the rich harvest of the west coast and the far less prolific grounds to the east, even if the eastern coastline is three times as long. Though the Baltic is not without its supply of herring, salmon and cod, in comparison to the Skagerrak, Kattegat and neighbouring North Sea, its resources are much less profuse. With the western regions of Bohusland, Halland and Skane within reasonable reach of these fish-rich areas, these waters were worked by a host of craft. On the other side of the country much fewer vessels fished. Those that did generally fished among the islands and rivers which indent the eastern coastline. Statistics illustrate this divide. In 1914 there were only fifty decked boats north of the Blekinge region in the east while some 1,774 similar boats worked from harbours and beaches along the western side; 1,130 of these alone belonged to the Gothenburg/Bohusland area.

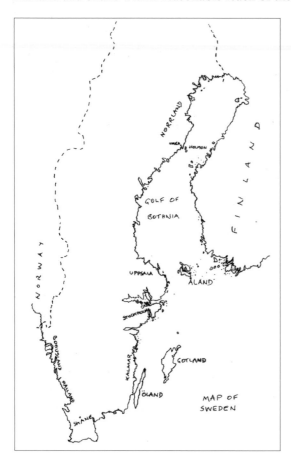

MAP OF SWEDEN

THE EARLY HERRING FISHERY

In 1124 there were vast shoals of herring to be had in the southern Baltic. By the end of that century an equally vast herring fishery had developed, this today being regarded as the first great herring fishery of Europe. Along the entire south-western coast, centred on the towns of Falserbo and Skanör, in what was then part of the Danish kingdom, thousands of small oared boats are said to have landed herring.

During the autumn months of September and October, the fishermen set up temporary camps along the shore. Here they cured the catch before sending it away in cargo ships. For over 300 years this fishery continued unhindered, and it is said that 7,515 vessels took part in 1494, although some dispute this number, which does seem excessively large. By this time, the north German merchants from towns such as Lübeck and Bremen had arrived, buying up all the catch, building their own curing stations, and exporting it in every direction. In other words, they took over the fishery and before long they controlled the fishermen too, a domination which contributed to the benefit of these merchants, and from out of this arose the Hanseatic League. The ensuing powerful political force thus created had far-reaching influences stretching far and wide throughout the European continent. The fishery was of such importance that Rome even decreed that work could be allowed on Sundays and other holy days, suggesting more than a tenuous link between the Roman Catholic Church and these Hanse merchants in the Middle Ages. These merchants had representatives in many corners of the globe, including London, where they owned warehouses on the site of what is now Cannon Street Station, then called the Steelyard. So reliable was the currency of herring that merchants in Britain demanded payment in 'pounds of easterlings' – later to become 'pounds sterling'. The Hanseatic League, it has been suggested, was an early attempt at setting up the first globalised world economy, many centuries before the current attempt. Further suggestions have linked the Hanseatic League and its political control to the Illuminati! However, like many other things in history, the political force of the Hanseatic League disappeared almost as quickly as it had arrived, and probably as fast as the herring's migration, which caused it to desert the shores. It has been shown that increased rainfall in the Baltic and exceptionally high tides caused the herring to migrate into the North Sea to spawn, where in time they stimulated the fishery there.

It is assumed that these small rowing craft were of Scandinavian appearance, probably about 25 feet in length and built to the same traditions. They were undoubtedly strong boats, crewed by between three and six men, and according to one source, they could carry between 20 and 40 tons of herring, although this sounds rather unconvincing for a vessel of that size. They used an early form of drift-nets or fixed nets to catch the fish. Annually, some 10,000 tons of herring were said to have been landed. However, it is not until the nineteenth century that clearer details of these Swedish fishing craft emerge. The boats had evolved by this time from early squaresailed Viking vessels into perfected craft, albeit with influences from other parts of the continent. The fishermen found that the boat tacked better if they brought the front of the sail back to the mast, making the yard into the sprit, and giving themselves a higher degree of control, which proved effective in the confined waters around the islands. These open double-enders were clinker-built in a heavy construction and appear to have been between 16 feet and 20 feet overall, which suggests a size reduction from the early craft.

THE INSHORE FISHERIES

In more recent times the Swedish inshore fishing industry existed at a level much nearer to subsistence. Prior to 1870 fishing was for those who normally worked the land, when that land was quiet. The west coast is wild and exposed, with little shelter existing anywhere upon the coast except behind islands. Moreover, the herring shoals had never been seen on the scale of medieval times. However, with a harbour-building programme in the next decade, this changed rapidly and full-time fishing spread throughout the western and southern coastal regions.

Various types of boats worked off these coasts. Perhaps the best known of these is the *snipa*, similar to the Danish *smakkejolle*, as will be discussed in a later chapter, which was used for all manner of fishing – small net fishing, lobstering and using a *ryssja* net for catching eels from the eighteenth century. Although rowed while working, a spritsail was set when journeying to and fro the fishing grounds. These craft exhibited strong Norse roots with curved stems and sternposts, resembling the older *færing*, and typically were 17–18 feet, clinker built with eight to ten planks and open. They had to be seaworthy to work these exposed waters and some fitted a detachable 'splashing board' above the last plank. The mast was unstayed to allow for lowering when riding to the nets, although around Kullaberg, in the northern part of the Øresund, the fishermen had no need to sail the short distance out because the water is deep close to the shore. As the fleets developed, *snipor* (plural) of 25–30 feet were common in the late nineteenth century, especially in the waters of the Øresund close to the rich fishing grounds shared with Denmark. By the 1940s, though, the use of such boats for fishing ceased and today many different designs of *snipor* sail around the islands for pleasure only. Many are engined, which they accommodated well, have no rig at all and are used as launches between the islands. Along the Bohusland coast the *snipa* was referred to as a *snacka*, although technically this term refers to a smaller boat.

Vassing were bigger vessels which were also open and were gaff rigged. They seem to have been used mostly to take trippers out from the crowded summer beaches in the 1880s, although many were used for fishing outside that season. After 1905 many were engined and their working use continued up to the Second World War.

The quest for mackerel and herring took the fishermen of Bohusland closer inshore, largely around the Skagerrak and Kattegat. By the 1870s these were decked vessels

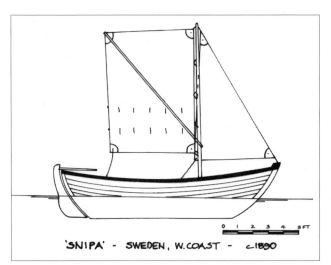

'SNIPA' - SWEDEN, W.COAST - c.1890

The multi-purpose *snipa* had strong links with the older *færing* type. The spritsail, although set here for the camera, was in its very basic form when this photo was taken, in about 1900.

Hamnen. Varberg.

A *snipa* in Varberg on the west coast, about 1900. This looks like it is regatta day, judging by the throngs of people. The helmsman has certainly not been out fishing!

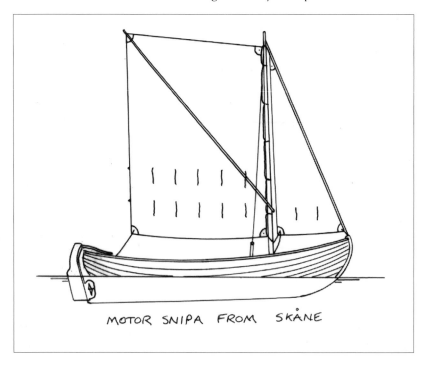

Below left: A *vassing*, which was a larger version of the *snipa* and was usually gaff rigged.

Below right: Another *vassing* at Varberg, possibly on the same regatta day. Many of these boats were fitted with engines in the first few decades of the twentieth century.

working a drag-net, hence the name of 'drag-net boats'. Also referred to as *garnbåt* (net boats), many originated from the Koster Islands up in the north, off Stromstad and close to the border with Norway. At 30 feet, they were similar in shape to the smaller beach boats, though finer below the water and beamier above. They were clinker-built and were speedy yet steady craft with plenty of capacity for fetching the catch ashore. Rigged originally with one spritsail when undecked, around 1850, they adopted the gaff rig and were half-decked to form a small cuddy with basic accommodation. Later, with a full deck, a jigger mast was added to some with a spritsail. These became known as the *koster boats* and, when not fishing, they would often be found working between the islands, carrying sand or building materials. The name was perhaps unfair, especially on the Bohusland coast, where many were built on Orust Island and many referred to the vessel as simply a *dacksbåt* (decked boat). Similar boats, with some degree more of curvature in the stem and sternpost, also worked further south off the Halland coast and into the Øresund, where they were sometimes called *Øresund kosters*.

One such vessel was the 35-foot '*Blixten*', built for John Anderson in 1868 by Jacob of Stamnas at Orust Island, but by 1938, although engined and with the addition of a wheelhouse, she had become a wreck near Gravarne. Luckily, her lines were taken off then by Olof Hasslof and Henry Magnusson of the Gothenburg Historical Museum. A clinker replica was built in the early 1990s by the *Allmoge Båtar* or Association of Old Folk's Boats and this boat, though now engineless, sails from the small island of Bassholmen. The same Association looks after a number of other traditional craft.

The design of the *koster boats* is said to have had a heavy influence on the design of the early motor fishing craft. Carvel-built craft, some with counter sterns like many smacks, they retained their main and mizzen gaffs and often landed their catch into Britain, becoming known as 'England-farers'.

Other small craft were in use on the west coast. A *robåd* from Bohusland was similar to the *snacka*, coming perhaps somewhere in size between it and a *snipa*. Being an ancient craft, it was originally rigged with two spritsails, a rig common at the time but which was retained here longer than most other areas. The obvious literal translation of 'rowing boat' is misleading because of the rig. Eventually the vessel adopted a single sprit with topsail and jib, grew in size to 23 feet, and had a small cuddy under the short foredeck. By the time

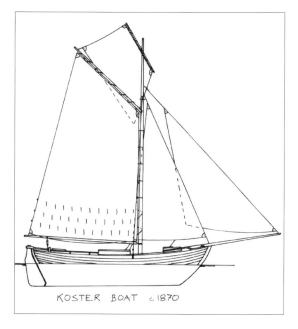

KOSTER BOAT c1870

A 1990s-built replica of the 35-foot '*Blixten*', built in 1868 by Jacob of Stamnas at Orust Island, under full sail in 2000. (*Photo: Hermann Ostermann*)

they were fitted with engines in the first couple of decades of the twentieth century, they were indistinguishable from the other craft of the west coast.

A *kag* was a small flat-bottomed boat for use in the inland waters. Some of these set a spritsail, though most worked under oars. Other flat-bottomed vessels were used for fishing in the shallow water, primarily for salmon, and another type of these was the *eka* or *laxeka* (*lax* = salmon). Those that worked off the exposed coast of Halland were unique in having an uplifted bow to help them ride the swell, this being called the *uppnosig*. In the north, among the islands, this uplifting wasn't needed and there are similarities between these and the salmon cobles of eastern Scotland.

Straight sides were desirable when working from the boat with the *bottengår* nets – a system of fixed nets set on poles in 10–15 feet of water. Other *eka* were clinker-built with the traditional profile shape, yet have a flat bottom and centreboard. These have a spritsail and jib and often sailed far offshore when fishing. Being flat-bottomed, they were easy to beach on the flat, sandy, open coast.

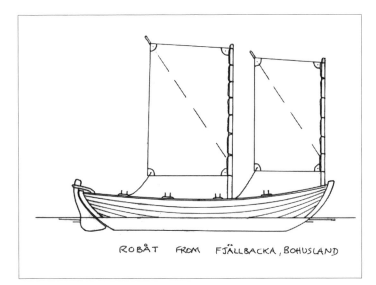

ROBÅT FROM FJÄLLBACKA, BOHUSLAND

Below: A flat-bottomed *eka*, with its upturned bow to ride the swell.

EAST COAST

Similarly shaped craft also worked on the east coast, though with some transom-sterned craft being favoured. Indeed, this sort of vessel was once prolific around the Baltic and could be found working off beaches in Poland, Lithuania, Latvia and Estonia. The east coast also has its own type of small open fishing boats which, with raking stems and transoms, resembled dugout canoes. These were in use among the islands and were easily beached in winter when the Baltic froze over.

In the Kalmar sound, single or double-masted *lotsbåten* worked the narrow waterway between the mainland and the island of Oland and many of these were transom-sterned. Gotland, further out into the Baltic, was an important fishing base where boats were single, double or triple-masted. These double-enders all set spritsails and the hulls had curved ends, a clinker construction and a distinct sheer line. The largest were some 26 feet overall.

Around the Stockholm archipelago and out to the Aland Islands, which are under Finnish control today, the spritsail was still the favoured rig. Indeed, it was around most of the Gulf of Bothnia and throughout most of Finland. One exception was the lug-rigged *skotbåtar* from around the coast north and south of Hudiksvall, in the south-eastern part of the Gulf of Bothnia, though this appears to have been a coastal peculiarity. This vessel had two masts, with the forward mast set well forward in the eyes of the vessel, which suggests it was a direct variant on the sprit-rigged craft whose masts are in a similar position. Also, in a country where clinker construction was the dominant mode of boatbuilding, one of the first motorised, carvel-built boats in the Swedish Baltic emerged from this part of the coast in about 1914. Generally, motorisation didn't have a huge effect on the small coastal communities until the 1920s.

Around Umeå and the island of Holmön, single or twin-masted spritsailed craft had a very basic rig and the boats were characterised by a steeply raked, straight sternpost. However, all along this coast, boats were generally light to enable them to be brought ashore once the sea froze over. Small, wide-planked, clinker-built boats, up to about 15 feet, worked the inshore rivers and estuaries while various types of flat-bottomed canoes were to be found in the rivers and lakes.

The east coast of Sweden is famed for its *stromming*, which is Baltic herring and must be mentioned. This is the main species sought, though other fish do get caught, especially salmon which, in the main, are caught in the rivers and estuaries. Crayfish is another delicacy. Soured herring – *surstromming* – is perhaps more of a national institution than a local delicacy, though for the not-faint-at-heart, for this is fermented herring that is left in a can for a number of months to mature (sometimes it is enjoyed when at least a year after it's out-of-date) and carefully opened once the can is bulging. The smell is said to permeate through walls. Much of the herring is caught in seine nets worked from the shore using one small boat.

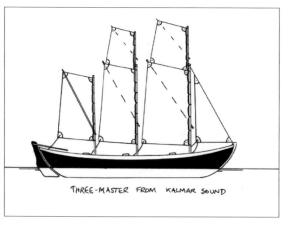

THREE-MASTER FROM KALMAR SOUND

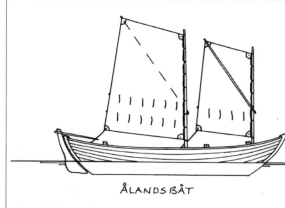

ÅLANDS BÅT

OFFSHORE CRAFT

Generally, two types of larger sailing fishing boats worked from the Swedish waters of Bohusland out into the North Sea. On deserting the inshore waters of the Kattegat and Oresund, the herring moved into deeper water where the fishermen soon made bigger catches. Documentary evidence of these early boats is non-existent, and it wasn't until the eighteenth century that some insight can be gained into these craft, mainly through the writings of priests.

The earliest type of documented craft worked the shallower banks of the North Sea, hence its name of *bankskutor* - literally '*banksboat*'. These heavy, oak-built half-deckers had two or three masts; when fishing, the forward of these was lowered to lie to the mizzen, unless strong winds prevented this. They set squaresails with a sprit mizzen, and the sails were normally woven from hemp by the fishermen themselves. After about 1860, gaff rig was adopted. It is not known when decked *bankskutors* first appeared but, when they did, the forecastle was enlarged and the hull sub-divided into compartments for gear and the catch. Boats at this time were up to 70 feet long on a 60-foot keel. The crew of six lived in the cramped space for up to six weeks at a time. They would only return home once the boat was full, unless they ran out of food before that. Each man kept his own supplies of bread and strong drink in a chest, and the cuddy had a stove to keep them all warm and for cooking. Next in importance, a barrel of water was kept carefully in the accommodation. In reaching the fishing grounds, the *bankskutor* was anchored (the anchor had a wooden stock) and the two *lillebat* ('small boats'), from which the lines were set, were launched. These were baited with 'blue mussels' collected before setting sail. However, these boats were particularly awkward to handle, which was why most fishermen had given up using them by the latter half of the nineteenth century. Those from Mollosund and Karingen persevered with them for a bit longer.

The *bankskutor* was superseded by the *sjöbåtar* – literally 'seaboat'. The earliest of these were clinker-built, often by the fishermen themselves. Timber was plentiful locally, and cheap, though expensive imported oak was used for the centreline. The local cut pine was obviously used as much as possible. At about 50–55 feet overall, these boats were smaller than the older boats, with a keel some 10 feet shorter. They only had two masts, the third having been abandoned with the phasing out of the *bankskutor*. Rigged with gaff mainsail, topsail,

A 'BANKSBOAT' IN HEAVY SEAS!
from a lithograph dated 1840 —

Lillebatar – literally small boats – at Gullholmen on the south-east coast.

staysail and jib, the *sjöbåtar* were much easier to manoeuvre, mainly due to their short, sharp and deep underwater body. Their tall, wide and rounded shape above the water enabled them, it is said, 'to ride the waves like a seabird'.

Carvel construction was adopted in the late nineteenth century, with oak trenails being used when available, otherwise 'pine-nails' or galvanised spikes. After 15–20 years, these boats were retired from fishing and used to move cargo over to Norway. The five crew of these vessels were part-owners and had equal shares. Each man had his own task at sea and he just got on with it. Often it was a 'father and sons' relationship. Like the older boats, they fished far out, venturing to Shetland and into the Atlantic. As before, they fished from spring to autumn, using one small boat to set lines or tow drag-nets. Once the small boat was fishing, the remaining crew often had impromptu regattas with other sea boats where speeds of up to 10 knots were not unusual. Fishing continued until late autumn, when the boats were taken ashore and scrubbed, painted and tarred.

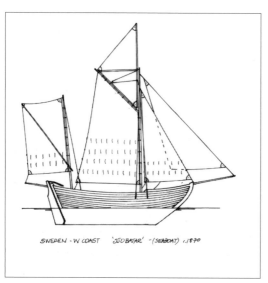

SWEDEN - W. COAST   'SJÖBÅTAR' - (SEABOAT) c.1870

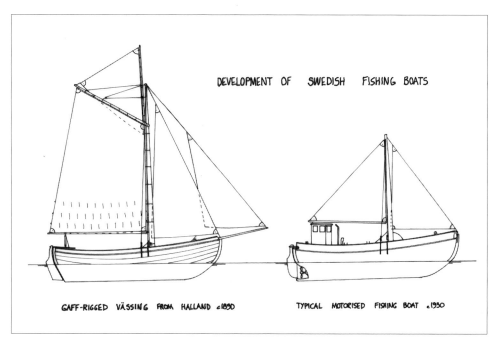

DEVELOPMENT OF SWEDISH FISHING BOATS

GAFF-RIGGED VÄSSING FROM HALLAND c1850

TYPICAL MOTORISED FISHING BOAT c1930

The small boats were very seaworthy clinker-built double-enders which were very beamy (2:1 length:beam ratio) and were able to carry a good load of fish. Spritsails were set to tow lines, while oars alone were used when drag-netting.

British smacks, locally referred to as 'cutters', were introduced into the offshore fleet towards the end of the nineteenth century. These Brixham and Lowestoft vessels had fished for cod and ling around Shetland and so had been observed by the Swedes. When steam began to have its impact and the British started selling off their smacks, the Swedes immediately bought those being withdrawn, helping to bring about the total transition from sprit to gaff. By the end of the century these smacks had just about taken over from the sea boats, so that the latter disappeared. Today, many British-built smacks still sail under the Swedish flag, chartering or for pleasure.

SOURCES

I've not discovered anything of substance written in English concerning Swedish or Finnish fishing craft. For Swedish traditional types, see *Allmogebåtar* by Alvar Zacke and Magnus Hagg. The National Maritime Museum is in Stockholm, concentrating mainly on the history of the navy and the warship *Vasa*. Göteborg (Gothenberg) has a small fisheries museum while the Brantevik Maritime Museum has some fisheries related artefacts. There's also a fishing and maritime museum at Raa, Helsingborg. The small museum *Båtmuseet Galtaback* has a most interesting collection of craft. The *Sveriges Segelfartygsförening* (the Swedish Sailing-Ship Association) has published at least two books on traditional craft.

CHAPTER 3

Finland

The Gulf of Bothnia to the Gulf of Finland

Finland is geographically different to most European countries in that although it has over 2,800 miles of coastline and more than 180,000 islands, half of these islands lie in the 188,000-plus lakes that cover the country. No wonder it is called 'the land of a thousand lakes', even if this is a huge understatement. It also has Europe's biggest archipelago – the Åland Islands, which consist alone of 6,500 islands – lying off its south-west coast. All of this suggests waters teeming with fish and a population relying upon a seafood diet. However, probably due to the long winters, with much of the coastline frozen over, it is only in the late spring to early autumn season when, as in Northern Sweden, salmon, Baltic herring and crayfish command many a menu, with enough being put aside to last through those dark months.

Finland has for centuries been ruled by either Sweden or Russia. Independence was only gained in 1917, becoming a republic two years later. It was at about the same time that the country gained control over a part of the country in the north known as Petsamo, a strip of land – or arm of Finland – reaching to the Arctic Ocean. Petsamo, during the period of Finnish control, was the country's only access to the rich fishing grounds of the north and

Seal boats at Porvoo, on the eastern end of the Gulf of Finland. Note the two-masted boat in the background while the high, flaring bow of two of the boats reflects that of the photo at the bottom of page 44. (*Photo: Finnish National Board of Antiquities*)

Small oared craft at Koivisto on the Gulf of Finland in 1930. Similar craft were found all over the inland lakes of Finland. (*Photo: Finnish National Board of Antiquities*)

was consequently an important fishing station until 1944, after which it was ceded to the Soviet Union after the Winter War of 1939–40. This was demanded by Stalin because, although the area was indigenously populated by the Sami, it had previously been part of Russia since 1533.

Finland's boatbuilding tradition lies in the Sami sewn boats brought down from the north, though several outside influences tended to meet in the south of the country. From the west came the Norse clinker boatbuilding techniques while across the Gulf of Finland, in Estonia, only some 30 miles across the sea, the central European dugout canoe brought its own influences to Finnish craft. However, by the nineteenth century it appears that many of the vessels mirrored those of the opposite side of the Gulf of Bothnia, and thus were wide-planked double-enders, mostly with square, and later sprit, rigs, though sewn boats and flat-bottomed canoes and dugouts did continue to work the inland rivers and multitude of lakes within the country. Prior to this, the majority of clinker-built vessels were purely trading vessels, generally referred to as *skutes*. Though these seldom fished, some were owned by fishermen and the so-called peasants who carried fish over to Estonia and Sweden. Until 1809, when Finland was no longer part of the Kingdom of Sweden, fishermen from the south-west had to carry their catches to Stockholm for selling. The boats they used were at first open boats rigged with a single square sail, though as they grew in size more masts were added so that by the nineteenth century some had three masts. The sails were usually woven fabric, which was superseded by cotton towards the end of the eighteenth century although the older material survived longer around the Åland archipelago and Turku. Carvel-built schooners became popular after the 1830s, when peasants were allowed to trade further afield. For the fishermen in the extreme east of the Gulf of Finland, in Karelia, most fish was sold in St Petersburg.

Gouged-out, flat-bottomed boats have been discovered dating back to the Stone Age. In the south and west of the country, planked flat-bottomed vessels were used for coastal fishing. Planked craft were normally of four planks each side, though in the eighteenth century, in Åland, some had five. Most fishing was either seine-netting from the shore with one boat, drift-net fishing for Baltic herring further out, or long-lining for cod. Transom-sterned seine-net boats also developed in the nineteenth century. However, before drift-netting was introduced by guides specially imported from Gotland to teach the locals in the 1860s, herring was largely caught with fixed nets, known locally as hooked nets. Use of the drift-net soon spread west along the Gulf of Finland and into the southern part of the Gulf of Bothnia. One particular fishing station of importance was on the small island of Suursaari, where a thriving trade in fish developed.

Only by studying photographic evidence from the nineteenth and twentieth centuries has it been possible to form some idea of the diversity of coastal boats in use in Finland. The seine boats appear in varying sizes all along the coast. Many of these were entirely undecked, clinker-built with up to eight planks over a length of some 20 feet. The northern boats from places such as Kemi show slightly wider planks than those from further south. One eight-planked vessel from Suursaari (which was part of Finland until 1940 and is now Russian territory and called Gogland) had two

The island of Suursaari once belonged to Finland and was an important fishing station until the Russians gained control over it in 1944. Today it is known widely as Gogland.

A typical boat at Suursaari in 1923. This clinker-built, seven-planked boat has two mast steps and is probably similar to that two-master in the middle photo on page 44. The one-piece frames are surely steamed, except for the three back from the bow. (*Photo: Finnish National Board of Antiquities*)

Above: Hauling in a seine-net near Turku in the south-west. Unusually, the net is being pulled from the two boats and not the shore. (*Photo: Rauma Maritime Museum*)

Left: A square-sailed boat at Naantali near Turku in the south-west. Note the long sweeps, used when setting the seine-net. (*Photo: Finnish National Board of Antiquities*)

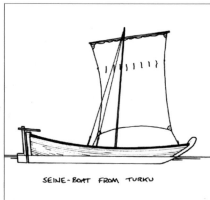

SEINE-BOAT FROM TURKU

thwarts with sawn frames that appear to meet over the keel so that there are no floors in typical fashion. All appear relatively light in construction, probably due to their being carried ashore in winter once the big freeze arrived, to be stored under cover when possible. Indeed, it seems that most houses near the coast had boathouses for this purpose. Often, boats were owned by Seine Associations, with equal shares paid out, though boat owners might receive more than simple fishermen.

One exceptionally unusual type of boat came from Raippaluoto on the Gulf of Bothnia, close to Vaasa. These had high, flaring bows resembling some modern-day Spanish west coast boats! The after end was spoon-shape, rounded rather like a small canoe. Length was judged to be about 30 feet and these were totally open.

Inshore planked canoes of some three planks were commonly owned by the peasants and were big enough to carry three men. Some of these set a small spritsail. Vessels such as these were often used to service the sometimes extensive fish weirs set into the water.

A typical small three-planked boat at Rauma, on the Gulf of Finland. (*Photo: Rauma Maritime Museum*)

Four boats used to service the numerous fish weirs. These were flat-bottomed craft, over 25 feet long and, though basic in build, were strong and efficient workboats. (*Photo: Finnish National Board of Antiquities*)

Above left: Two seine-net boats, each with seven crew, being towed to the fishing grounds at speed, presumably by a motorised vessel. (*Photo: Rauma Maritime Museum*)

Left: A two-masted sprit-rigged boat from Rauma, posed for the camera on a very calm day with all the family aboard! This type of boat was typical of the south and south-west coasts and over to the Åland Islands. (*Photo: Rauma Maritime Museum*)

Below: Boats from the Gulf of Bothnia, at Raippaluota. (*Photo: Finnish National Board of Antiquities*)

According to Ilmar Talve, there were some fourteen different main types of boat throughout Finland such as the *Kokemaenjoki boat*, the *Pyhajarvi-Koylionjarvi boat* from the southern part of the Gulf of Bothnia, the *Finland Proper boats* from the south-east and those from the Gulf of Finland and the north. Although most fall into one of the above categories, regional variations came into play, as did individual traditions of the fishing associations and of course the boatbuilders themselves.

As elsewhere, motorisation affected the type of boats that were able to receive small inboard units. For many areas, this didn't happen until after the Second World War, subsequent to when Finland fought the Winter War with Russia and miraculously (some say) held back the might of the Red Army, only losing small amounts of territory, such as in Karelia and Petsamo, the latter being where, as already mentioned, its access to the rich fishing grounds of the Arctic was lost. Previously steamers, and

Above: A counter-sterned motorised cutter-type boat at Rauma in about 1930. (*Photo: Rauma Maritime Museum*)

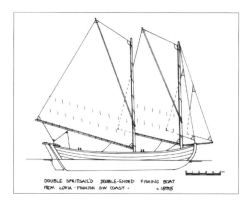

DOUBLE SPRITSAIL'D DOUBLE-ENDED FISHING BOAT FROM LOVIA - FINNISH SW COAST - c.1895

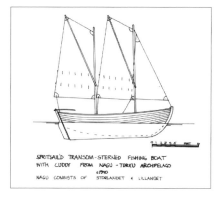

SPRITSAIL'D TRANSOM-STERNED FISHING BOAT WITH CUDDY FROM NAGU - TURKU ARCHIPELAGO c.1910 NAGU CONSISTS OF STORLANDET & LILLANDET

later motor boats, had landed at Petsamo. After 1944 all fish either had to come from the Baltic or be imported.

Since that time fishing has still played an important part in the economy and the design of motorised vessels followed a typical Baltic pattern of wooden cruiser-sterned boats with apple-cheeked bows, smaller yet similar to Danish and Swedish craft. Those that remain are mostly converted to pleasure these days, while examples of the older, traditional, clinker-built boats are hard to come by. Those that do are normally ashore as exhibits or rotting remains. Some replica vessels have, thankfully, been built.

SOURCES

For Finnish traditional types, see Ilmar Talve's *Finnish Folk Culture*. Information concerning Finnish types also came from the Finnish National Board of Antiquities and the Rauma Maritime Museum. The National Maritime Museum is in a new location at Kotka. The small museum at Rauma has a photo archive.

A typical Finnish boatyard. (*Photo: Rauma Maritime Museum*)

CHAPTER 4

The Baltic States:
Estonia, Latvia & Lithuania

The Gulf of Finland to the Curonian Lagoon

ESTONIA

Of all the countries bordering the Baltic Sea, it is said that Estonia has the most varied coastline: a mixture of rocky bays and sandy beaches backed by dunes. It also has somewhere in the region of 1,500 islands scattered around the north and west shorelines, the largest two being Saaremaa and Hiiumaa, both home to fishing communities which largely survived at a subsistence level. Compared to the other two Baltic states of Latvia and Lithuania, it has a far greater length of coastline, at around 2,352 miles.

Between approximately the thirteenth and twentieth centuries Estonia's history was one of a continual struggle against invaders – Swedish, Danish, German and Russian – all of whom contributed to the type of boats in use. Primarily, it seems that the Swedish occupation was regarded as the most tolerable in what was sometimes referred to as the 'golden era' and which brought about the main influence on the design and construction of fishing craft. German and Russian interference, especially in the twentieth century, resulted in the destruction of the vast majority of traditional vessels. Russia also regarded

Two typical Estonian boats on the beach on the coast of the Gulf of Finland. A spritsail is set upon the short mast. Note the wooden logs to enable the boat to be dragged up the beach, although there is almost no tide in the Gulf.

A two-masted *jaala*, the sprit-rigged boat of the north and west, coming ashore somewhere near Tallinn.

all three Baltic states as a source of food for the Soviet empire, though this was mostly concentrated upon agriculture. Fishing was discouraged for fear of the boats fleeing and what fish was landed was quickly sent east when possible. Even up to the twentieth century, the fishing rights up to three kilometres from the coast were owned by the lords of the manor.

As elsewhere about the Baltic, it was the herring and sprat that were the most sought-after fish, followed by cod and salmon, while eels, smelts, sea-trout and lampreys were fished during their migration upriver. Even before the era of the Soviet Union Russian fishermen were controlling the fishing and they introduced big seines. Finnish fishermen brought in net traps for the herring. Lake Peipsi, Europe's fifth-largest lake, which borders Russia, was home to a substantial fishery where today some 100,000 tons of fish are extracted. Almost forty different species are to be found there and methods as varied as 'burbot tackle', 'noose fishing', fish-forks, kiddle nets and landing nets were in use. Ice fishing occurred in winter and traditional boats were small canoes called *lootsik* or *kuna*, or larger *vene*, a characteristic boat of the lake with the stern part ending with an abrupt and upright *tohupakk*, made of a split log. In the fore end of the vessel there is a peculiar upright and curved stem. In between, the boat was carvel-built with a few frames. For moving fish and other goods about the lake and the tributary rivers, especially the River Emajogi, and trading with the people of Novgorod (part of today's Russia), the *lodi* was the typical barge – sometimes referred to as the Hanseatic barge – a replica of which has recently been built. Although these were once prolific on the lake – 200 were said to be moored at Tartu (once a member of the Hanseatic League) on the River Emajogi at any one time – none survived, especially after the ravages of the Second World War, when they were found to be suitable landing craft and were thus soon destroyed. This replica, the 12-metre clinker-built boat '*Jommu*', is covered over by a raised deck and is rigged with one huge squaresail. Owned by the Emajoe Barge Society and crewed by volunteers, she sails around the northern Baltic in part of the summer while operating trips from Tartu at other times, though obviously not in winter when the waters are frozen!

The *jaala* was a two-masted, sprit-rigged boat found on both the north and west coasts and was characterised by its main mast being placed centrally. Lengths were approximately, on average, 10 metres. These vessels, mirroring those from across the water in both Sweden and Finland, have evolved from Swedish craft over the centuries from the Swedish communities that settled in the area, it is said, over a thousand years ago. Later versions rigged a gaff sail on the higher mainmast while retaining the sprit on the foremast and setting a jib. These boats, some with a transom stern, reached their peak between about 1860 and 1920, when many were used exclusively for seal hunting. Today a replica, built in 2003, is based on the small island of Ruhnu, in the Gulf of Riga, and is used for day-trips and longer packages to take people out to the islands of the Vainameri Sea. The people of Ruhnu only spoke Swedish until 1944, when 300 of them escaped back to Sweden, leaving behind only two families. Traditionally, the Estonian Swedes would sail these boats back to Sweden to claim

Fishermen's canoes near Kallaste on the shores of Lake Peipsi in 1924.

Above left: The typical *lodi* barge of Lake Peipsi was used to transport everything, including fish.

Above right: A small gaff-rigged boat similar to the *jaala*, displaying some Swedish characteristics.

their rights and so, when the replica was launched, she was sailed back to deliver a letter to the King of Sweden.

In Lake Vortsjarv, although not a sea boat, the *kale* exhibits a strong resemblance to many other Baltic craft. The *kale* is a 12-15-metre clinker-built trawler with a sprit-rigged main and jib that towed a net of the same name. Between the 1930s and the 1970s there were seventy *kaleships* on the lake, though, due to a change in fishing methods, the Soviets finally destroyed all of them. Again a replica, the 12-metre 'Paula', was built in 2005 under the guidance of the only remaining shipwright who had learned the skill in his childhood and was able to pass this on to local fishermen from the village of Valma.

LATVIA

Latvia's coastline is much shorter at some 330 miles and is one largely of sandy beaches where boats would have been pulled up when they were not exploiting the rich resources of fish close to its shore. Like the Estonian cities of Tallinn and Tartu, Riga became a member of the Hanseatic League in the mid-thirteenth century, though by that time it was already a trading centre where furs, herring and spices arrived inward while exports of linen, wax and grain were made. Fishing, despite being well developed along the coast and with fish being a major part of the diet, remained at a subsistence level right up to the early twentieth century and continued to be restricted to the coast of the Baltic until when, in 1954, fishing out into the Atlantic began for the first time. Again, Russian influence and discrimination prevented a large-scale fishery from developing.

However, with a similar upheaval against aggressors from outside and especially Sweden again bringing influence through a cross-flow of the population, the traditional vessels used by the littoral fishers appear to have been similar in shape and construction to the *jaala*, as mentioned above. Most were clinker-built, tarred inside and out, and sprit-rigged.

LITHUANIA

Lithuania has one of the shortest coastlines in Europe at 62 miles, even though it is the largest and most populous of the three Baltic states. Almost half of this coastline is made up of the impressive Curonian Spit, a narrow strip of land separating the Curonian Lagoon from the sea. This it shares with neighbouring Kaliningrad, though the only exit to the sea is in Lithuanian waters at Klaipeda. This spit, one of three in the southern parts of the Baltic – the others being the Vistula Peninsula (Kaliningrad controls the outlet to the sea) and the Hel Peninsula that forms the Bay of Puck – is made, as are the others, by the action of the current in moving sand along the coast, thus forming the narrow, sandy peninsulas that are still 'growing' in length today.

Other than small flat-bottomed canoes which are similar to those from Poland, which we will discuss later (much of today's Poland and Lithuania were, in the sixteenth century, linked as the Republic of Two Nations, one of Europe's largest Western powers), the indigenous work boat of Lithuania is the *kurenai* (kurrenkahn), the flat-bottomed vessel of the lagoon used primarily for fishing, though also for carrying hay, cattle and other goods and even for going to church. Their name, however, comes from the type of net they used, a three-walled net (*kornas*) that they dragged in pairs downwind. A larger version, referred to as a *kuidelvalté* (keitelkahn), at up to 14 metres and usually gaff-rigged, drifted with a *kiudelis* (trammel) net, with the name of the vessel again coming from the type of net. Thus a smaller sprit-rigged boat setting a *kiudelis* net is called a *kiudelvalté* but, for the ease of understanding, we shall refer to the *kurenai* as the smaller of the type.

These *kurenas* (plural) have been traced back as far as the time of the Teutonic Order in the fourteenth century. They ranged in length between about eight and eleven metres and were very beamy (up to 4 metres) and heavy, weighing typically 4–5 tons. They had one or, more normally, two masts and were rigged, as mentioned, either with spritsails or gaff, depending on which side of the lagoon they were from. The second mast is interesting and apparently unique in that it is extremely small compared to the main mast – approximately one third of its length – and sits just in front of the latter, against the bulkhead of the foc'sle. This always carries a spritsail even when the main is gaff. The fishermen of the Curonian Spit favoured the sprit while those from the eastern shore of the lagoon generally preferred the gaff. Most also set a jib (or *horn*) and furthermore, when they were drifting to their nets, a trapezoidal sail called a *brumas* was set. All these sails were made from either flax or cotton.

It is said that these craft were usually built at the fisherman's homestead by shipwrights from parts of the south coast of the lagoon, inside today's Russian Federation province of Kaliningrad. They were built in an amazing three or four weeks by one shipwright, his assistant and one or two apprentices. The bottom planking was built with pine of 10–12 cm thickness while wide planks of oak were used for the sides, strengthened with oak grown frames. All had leeboards because of their shallow draught of 30–40 cm, vital in a seaway with an average depth of 3.8 metres and widespread shoals. They were crewed by between 3 and 5 men and stayed out fishing for two or three days; the men had basic accommodation aboard at both ends and cooked on a fire made on gravel or crushed tiles spread over the bottom of the boat. It has been estimated that in 1939 there were about 250 of these larger fishing boats in the lagoon and up until 1945 fishing was only allowed under sail. After the end of the Second World War, when the Soviet Union ruled the area, the local population was either forced to leave or deported, with new populations being brought in from other parts of the Soviet Union. The boats left behind in the Kaliningrad region were cut up and used mostly for firewood, though some boats did remain at work in the Lithuanian part of the lagoon. However, by the end of the 1960s these had been replaced by motorised vessels and most of the surviving *kurenas* disappeared.

Above left: Gilge was one of the main ports of the fishing craft of the lagoon, although now it's in Kaliningrad. Here, a larger gaff-rigged *kuidelvalté* is in calm conditions.

Above right: The smaller *kurenai* was similar in shape and both types set leeboards.

A *kuidelvalté* on the river at Gilge showing the *kiudelis*, the trawl-net, dragged between two boats.

Left: A good deck view of a *kuidelvalté*, showing the layout. Note the hefty cross-beam just forward of amidships and the net on the port side.

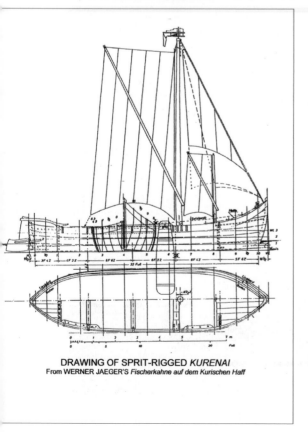

DRAWING OF SPRIT-RIGGED *KURENAI*
From WERNER JAEGER'S *Fischerkahne auf dem Kurischen Haff*

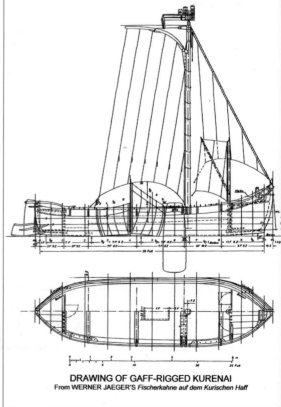

DRAWING OF GAFF-RIGGED *KURENAI*
From WERNER JAEGER'S *Fischerkahne auf dem Kurischen Haff*

Above left: A peaceful lagoon-side scene with the nets drying on the yard of a *kurenai*.

Above right: Here, at least ten *kurenas* are moored to the shore, sails holding them against the beach and anchor lines preventing complete grounding.

A picture of a gaff-rigged *kurenai* dated 1881. The design didn't change over many generations.

In 1989 the Lithuanian Sea Museum restored one remaining vessel, which had been built in Nida in 1935 and registered as NID.1, and this *kurenai* was launched the following year. For ten years the boat sailed around the lagoon every summer, taking visitors to regattas and festivals, until it was put ashore as a static exhibit at the Museum on the northern tip of the spit. Since then, two replicas have been built and launched at both Nida and Klaipeda and a further one in the museum is now used to promote awareness of the maritime heritage of the region. The other two still take visitors out and around the lagoon.

One distinguishing feature of the *kurenas* was the weathervanes on the tops of the masts, which they were forced to hoist after the fishery officers, in 1844, insisted that all vessels fishing in the lagoon do so; they also had to be marked with their respective village markings. The sails and sides of the boat also had to be marked with the letters of their village or the fishing permit number. Rectangular shapes formed the weather vanes, which were then painted either black and white (Curonian Spit villages), white and red (eastern lagoon villages) or yellow and blue (southern lagoon villages). Before long the fishermen began decorating these with refined carvings and thus they became symbols of the Curonian Lagoon.

Another inshore boat was the *valte*, typically 8–10 metres in length, which fished with nets and lines in spring and autumn, though they tended to seek colder water some 12–20 miles offshore during the summer. These craft consisted of five planks and were similar to a dory in shape, though they have been deemed Scandinavian in type. Most had a centreboard held in place with wedges. The *ketelhaven* is deemed to be similar to the Polish *Barkasow* (see next chapter). Built on a pine bottom with oak sides, this type is typical of fishing craft in use throughout the south and east Baltic coasts. The bottom might be replaced three times during the boat's career, with the original sides kept in place. Generally known as a very stable vessel with a 40–50 square metre sail area, only one is known to have capsized.

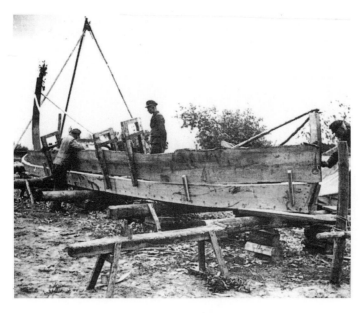

Building a *kurenai* in 1940. The bottom is completely flat – no rocker – and the planking has been cut by hand.

Above: An internal view of the flat bottom with massive frames and cross-beam, all of which are pegged together.

Right: Fisherman showing his model *kurenai*, complete with weathervane. (*Photo: Adrian Osler*)

The other smaller boats of the lagoon, all flat-bottomed and again named after the types of fishing, are the drag-net boat, the fish-trap boat and sailing boat and the *heuer*, a type, as we shall see, in use throughout the south and east coasts of the Baltic.

SOURCES

For the craft of the Curonian Lagoon, Werner Jaeger's *Fischerkahne auf dem Kurischen Haff* seems to be the main written source although Romaldas Adomavicius, the historian from the Lithuanian Sea Museum, has written several articles concerning the replica vessels (see 'The Return of the Kurenas. Sailing Boats of the Fishermen of the Curonian Lagoon' in *Baltic Sea Identity: Common Sea - Common Culture?*, editor Jersey Litwin) as well as helping the author in compiling part of this chapter. *Inshore Fishing Craft of the Southern Baltic from Holstein to Curonia* by Wolfgang Rudolph is tMonograph No. 14 from the National Maritime Museum, London. I found very little concerning Estonia, except from Lake Peipsi, although there is a maritime museum in Tallinn, and no written sources on Latvia.

CHAPTER 5

Poland

The Curonian Lagoon to the Mouth of the River Odra

In contrast to the Baltic states, Polish boatbuilders have not built many replicas of their traditional fishing craft – they don't have to as there are several collections dotted around the coast. Between the Fisheries Museum in Hel, the Vistula Lagoon Museum at Kały Rybackie and the Vistula River Museum at Tczew they have almost sixty traditional wooden vessels on display, the vast majority being used for fishing. Fishing was as vital to coastal, lagoon and river dwellers here as elsewhere and one look at a map of beach landings shows this in terms of forty-six places where fishing was practised, most of these being on the Baltic coast while several lie on the lagoons of the rivers Vistula and Oder (Szczecin Lagoon). Of those remaining, only eight were based in man-made harbours, though these were only built in the late nineteenth century.

Poland's Baltic coast is largely made of sandy beaches with dunes backing the foreshore and, in length, is similar to that of Latvia (305 miles). As mentioned previously, two peninsulas make up a small percentage of this coastline, the Hel Peninsula acting as a protector to the Bay of Puck, where the industrial centres of Gdynia and Gdansk (formerly the German town Danzig) lie, while the Vistula Spit almost cuts off the Vistula Lagoon from the sea with the only exit through the Russian waters of Kaliningrad. However, as the spit is only a mile or so wide, boats could be kept on both shores, depending on which was chosen to fish from. Thus at Kały Rybackie, for example, boats such as the trawling *Barkasow* would fish the lagoon while the typical *lodz rybacka* (fishing boat) would work the Baltic.

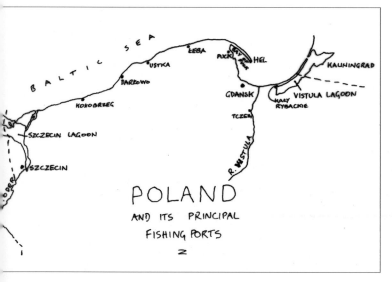

Fishing in Poland certainly dates back to the Bronze Age and since then various influences have played their part in the development of boatbuilding and design. The oldest plank boat found in Slavic lands was discovered in the Szczecin Lagoon, dating from the ninth century, after the Wends – Baltic Slavs who inhabited the coast from Holstein, Germany, to the mouth of the Vistula river – developed their fishing methods. Excavations have shown that the Wends living on the Gulf of Gdansk were fishermen. Another independent community were the Slovines, who became separated from other Kashubian Slavs after the partitioning of Poland in the late eighteenth century by Russia, Prussia and Austria. These Slovines ended up living on the Pomeranian coast near lakes Lebsko and Gardno and so they developed their own boat traditions. Considering the upheavals Poland has undergone over several centuries, it's hardly surprising that there's a wealth of different influences that have contributed to fishing craft. Given the subjugation of the country by the Soviet Union over four decades, it's a wonder that so many examples of fishing craft have survived. That they have in such numbers leads to one of two conclusions. Either the boats were still in use when democracy was achieved in 1989 or the Polish people were extremely keen to preserve and present their maritime history. Even though motorisation didn't have much effect on the beach-based fishing fleets until 1950, we know the former isn't true.

The rivers and lagoons were home to a diversity of flat-bottomed canoes, which have been classified by Wolfgang Rudolph depending on whether they were, as he put it, a bottom shell boat, a longitudinal-laid bottom plank boat, a thwartships-laid bottom plank boat or a keel boat. The first group are basically extensions of log boats whereby the hulls are built up from hollowed-out lumps of timber which act as the bottom of the boat. The second and third, more prolific, are typical flat-bottomed boats, though Rudolph identifies several ingenious methods of building their stem and sterns. The fourth, more akin to sea-going vessels, are more or less the same construction as Polish beach craft.

Canoes were in widespread use among the inland lagoons, lakes and waterways, these long, sleek canoes on the Weichsel estuary near Gdansk being used for transporting goods as well as fishing.

Two canoes on the Vistula Lagoon, the construction of both being very different. The method of attaching the frames on the right-hand vessel is particularly interesting as they are only fixed to the upper plank.

THE VISTULA LAGOON

Larger *Zakowka* are sprit-rigged vessels which fished for eels with trap-nets, and other fish as well, and became the most popular craft on the Vistula Lagoon. The method of construction has been said to have come from the nearby Danish island of Bornholm and was an adaptation of the traditional Viking method of clinker boatbuilding, belonging to Rudolph's longitudinal-laid bottom plank group. Here a flat plank (*czilplonka*), tapering at either end, which was sometimes bent using fire and weighted stones, formed the basis of the boat. Once it was shaped, it was cooled down with copious amounts of cold water. To this the stem and sternpost were butt-jointed, and successive planks then fitted in the reverse clinker method. The first – similar to the garboard – had an extreme twist, from flat along most of its length to the vertical where it met the stem and sternpost. Some boatbuilders used moulds though many, as elsewhere, simply used their eye and years of gleaned experience to produce exquisite shapes. Clenched-over nails were used throughout, a technique said to have been passed down from the *cog* builders of previous centuries. These boats had wet-wells and centreboards which were fitted through the wet-wells. Lengths were in the region of 5–6 metres and they were built mostly of oak though the bottom, which often consisted of one very wide plank, was pine. The sides were built up from three similarly wide planks. Most were rigged with two spritsails. Sometimes these craft were referred to as the *Zak* boats.

The *Barkasow* (plural of *Barkas*) range from 9–11 metres and were rigged with squaresails, probably the last square-sailed boats to have fished in European waters. A smaller version was the *Powbarkas*, while the biggest was the *Kaitel*, some of which had two masts and some of which were 15 metres in length. These boats worked in pairs trawling for white fish and eels at different times of the year. The net – also a *kaitel* – is said to have been recorded as far back as 1302, when the friars of Elblag were granted permission for *kaitel* fishing. The inhabitants of the shores of the lagoon

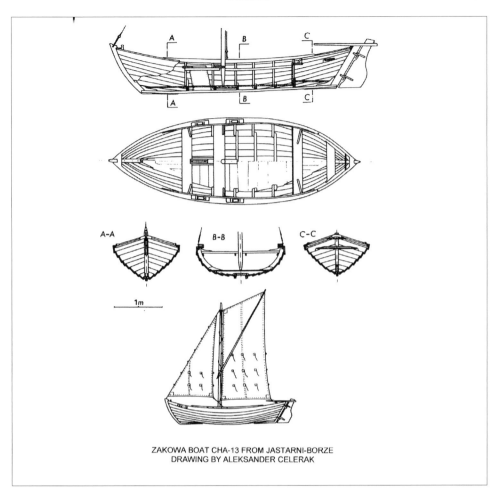

ZAKOWA BOAT CHA-13 FROM JASTARNI-BORZE
DRAWING BY ALEKSANDER CELERAK

A *zakowa* on the Vistula Lagoon. These craft are said to have originated on the Danish island of Bornholm.

Two *powbarkasow*, the smallest of the pair trawlers of the Vistula Lagoon, the last square-sail boats in Europe. These weren't motorised until the 1950s.

over the centuries, the Prussians, Germans and Poles, had all practised this type of fishing and thus the tradition had been handed down through generations. Post-1945, fishing continued up to 1971 though engines had, since 1955, been installed. However, only one boat of the pair was able, according to local regulations, to have an engine, which could only be used for getting to the fishing grounds. As most fishermen believed a motor scared away the fish, the legislation was largely pointless anyway!

The boats were built upon a pretty massive keel which was deeper at the bow than the stern. They were clinker-built in oak and pine, had wet-wells and were decked over with very basic accommodation for between four and six fishermen in the forward end of some of the vessels and net storage aft. When motors were introduced, these took up the storage space. Some of these *Barkasow* also had leeboards to counter their shallow draught of one metre, for the lagoon was only 2.5 metres deep in the western end and 3.5 in the east. However, with the seabed being flat and sandy, though covered in a layer of slime, it was perfect for trawling. When the same method was attempted in the Bay of Puck, the shoals ruined the nets.

Fishing was obviously impossible once the lagoon had frozen over though sometimes it was profitable to fish just before and after the freeze. Trawling for bream, pike and perch started with the thaw until May, then eel fishing May to September with white fish again from September to the winter freeze. During the winter all the fishermen could do was knock a hole in the ice and drop a hook or a net into the water. For this, they had sledges for carrying their gear back and forth, and taking the catch home.

The fishing method of what is basically pair-trawling was a tedious and complicated task. Two nets would be kept aboard, one for the white fish and a finer mesh for eels, though the fishing method was the same. Once they had reached the fishing grounds the two *Barkasow* would come together, facing downwind, and, using a strop and a wedge, would moor alongside each other, the hull being strengthened at this point. The net would be let out, which involved stitching two parts, one from each boat, together. Before 1945 this was up to 220 metres long by 11 metres deep, though subsequent to that date the Poles used a larger one, at up to 300 metres in length. To shoot the net the boats parted and sailed, one to the left and the other to the right, with the net spreading out between them. Once out, the boats would follow the same course and they would trawl for half an hour before the boats came back together to heave the net in. This had to be done quickly to avoid the fish swimming out, though engines, used illegally, were later found to keep the catch in the net. The net would then have to be unstitched once the fish was landed. Boats were built specifically to haul on the left or right so there was a slight difference in the layout of both. Sometimes they made as many as twenty-four hauls in one day. Four tons was an average catch.

The squaresail was found to be perfect for this mode of fishing downwind and, surprisingly, they performed well into the wind for the journey home. Kały Rybackie was the chief fishing harbour on the lagoon in the Polish sector. In 1948 there were eighteen pairs working while another few pairs worked from other harbours, including Peyse and Fischhausen in Kalingrad, though little is known about their history in this closely guarded and isolated part of the lagoon.

THE SZCZECIN LAGOON

Two boats worthy of mention that worked on this lagoon were the *hojer* and the *warpienka*. The former was a simple flat-bottomed canoe made up of one bottom plank and two side planks and is the same as the *heuer* from other parts of the Baltic. The *warpienka*, on the other hand, was an altogether fuller and heavier vessel, again flat-bottomed, though built up with four or more planks in a reverse clinker way. Up river, the *szkutka* was a traditional fishing boat of a canoe-style, partially decked over with a wet-well. Various other small canoes worked other rivers, Poland having one of the most diverse collection of river canoes in use over the last few hundred years of any other European country.

THE POMERANIAN FISHING BOAT (*POMERANKA*)

Working off the sandy beaches, the fishermen needed vessels that could land through the surf and these pioneered the flat-bottom construction (as described above) that most Polish fishing craft – except the canoes – developed. They originated from the western Pomeranian coast, where they were around 8.5 metres in length. Sometimes known as the *Dziwnow (Dievenowe) fishing boat*, or even *flounder boat*, they were generally used for drift-net fishing and drifting with hooks to catch flounders, salmon and cod. Most of these boats were owned by cooperatives of some three or more fishers who shared the cost and prepared the tackle together and then equally shared out the catches. A specific Kushubian fishing organisation with the same equal shares was the *maszoperie*, in which all the tackle was marked and owned by the organisation and the profits spread equally. By 1870 the use of these vessels had spread eastwards towards the Gdansk Pomerania, where they superseded older types. Many were fitted with centreboards to counter their shallow draft, though leeboards were more common in the nineteenth century before the adoption of these centreboards. Originally the boats were sprit-rigged, though many adopted the loose-footed gaff while a few even favoured the lug. What was vital as they came ashore was to be able to lower the mast quickly before hitting the surf and thus the mast didn't have shrouds but rope *karnatle* which could accommodate this. Beach capstans aided the hauling up process. Between the two world wars, two different variants of the Pomeranian fishing boat developed, one for the exposed beaches and a more heavily built boat in use in the sheltered waters of the Bay of Puck and the Gulf of Gdansk. Other variations of the Pomeranian fishing boat were the *Laskorn* boat of about 7 metres, the *Ceza* boat of 4.5–5 metres – a boat for operating seine nets – and the *Kon* boat of 4–5 metres. Other boats working from the Pomeranian shore will be mentioned in the next chapter.

A typical Pomeranian beach scene in the mid-twentieth century. Here the women are preparing the boat for sea, as well as mending the nets.

Boats were generally pulled out of the water using wooden capstans, the fishermen's wives always being part of the whole operation of fishing.

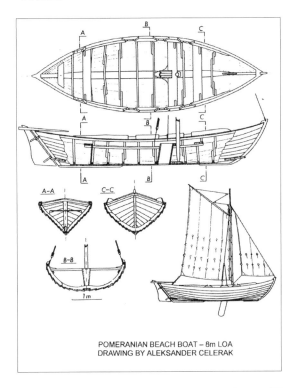

POMERANIAN BEACH BOAT – 8m LOA
DRAWING BY ALEKSANDER CELERAK

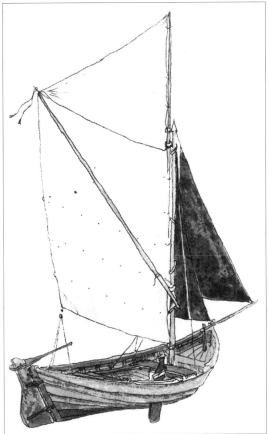

A wonderful water-colour of a Pomeranian beach boat.

THE DRIFT-NET BOAT

Introduced from Scandinavia, the round-ended drift-net boat was built upon a keel with clinker planking and mirrored Danish types. These first appeared on the Pomeranian shores in 1880. They were in the region of 6.5 metres in length and decked over and rigged with a single spritsail. In time they grew in length – up to 11 metres – and adopted the gaff rig. In the middle of the twentieth century, engines began to replace the rig.

THE CUTTER

Again based on Scandinavian designs, the first Polish cutter was built by Gdynia boatbuilder Franciszek Ledke in 1922 for a Gdynia fisherman. At that time there were twenty-five fishing villages in Poland, employing 1,206 fishermen with sixty-one motorised cutters (all presumably brought in), thirteen sailing drift-net boats, four other

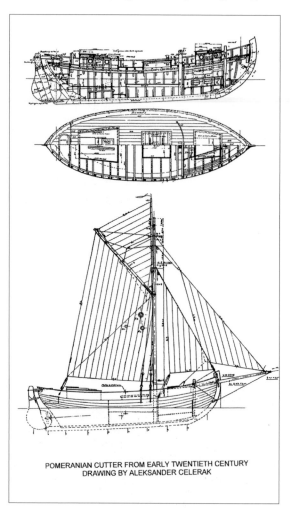

POMERANIAN CUTTER FROM EARLY TWENTIETH CENTURY
DRAWING BY ALEKSANDER CELERAK

motor boats and 841 sailing boats excluding, of course, the river and lagoon craft. By 1930 Ledke had moved to the Hel Peninsula and between 1931 and 1935 he built sixteen cutters, all characterized by their curved ends, wheelhouses and trawl gear. This yard was eventually transformed into a nationalised Fishing Shipyard. Many of these cutters fished up until the German occupation of 1939 but Polish fishermen were deported to concentration camps once the Germans had occupied, especially, the Hel Peninsula. German fishermen, who had been previously deported to the Reich by the Polish authorities two years earlier, returned with their own cutters while other German North Sea trawlers arrived at the same time. Cutters were sunk to block ports and the Bay of Puck during the German offensive but once Poland was occupied, local fishermen were allowed to recommence fishing but only during daylight hours and only under German supervision. This applied both to the open beach boats and the Baltic cutters. In 1945 there were 337 row boats and thirty-four cutters based in the Gulf of Gdansk.

ZUGA

The *zuga* was another flat-bottomed vessel that was primarily used to transport fish. It had a large wet-well amidships and a fore and aft deck. At around 10 metres, they were rigged with one boomed gaff mainsail and a small jib and were fitted with leeboards. Later versions had a centreboard through the wet-well. Their use was phased out in the 1930s as road vehicles enabled fish to be moved about quicker. Their use seems to have been mostly about the Gulf of Gdansk, taking fish from the villages of the Hel Peninsula to the towns of Gdansk and Gdynia. One such vessel, built by Juliusz Struck of Jastarnia and now on display in his museum boatyard in the village, was used to transport beer on the return journey. Juliusz is a fourth-generation boatbuilder and his predecessors have built fishing craft that are still in use.

MOTORISATION

As we've seen above, motorisation didn't have much effect until after the Second World War though some steam vessels had been brought into the Polish fishing fleet. Today the main fishing harbours are at Swinoujscie, Ustka, Wladysławowo and Gdansk, though there are harbours at Kolobrzeg, Darłowo, Kuznica, Jastarnia, Puck and Hel. Beach boats still work off a few beaches but the main harbours are used by large deep sea vessels. The smaller vessels today resemble the Scandinavian hull shape with curved ends though some have still retained the flat-bottom construction. Many are painted yellow above the waterline while others have small wheelhouses.

Motorisation had its effect in the 1920s, although the boats still had to work off the beaches.

Tourism also affected the fishing and here a beach boat is taking trippers out from a busy beach.

A motorised cutter, based on Scandinavian designs, at Hel. The church in the background is home to the Polish Fisheries Museum. (*Photo: Aleksander Celarek*)

Right: Two traditional boats on static display between beach and road half-way along the Hel Peninsula.

Below: Exhibits inside the museum include a square-sailed canoe and a Pomeranian beach boat.

SOURCES

The Polish Fisheries Museum at Hel has a fantastic collection of traditional vessels as well as a wealth of information on the history of fishing and a useful guide book (in Polish). Likewise, the Vistula Lagoon Museum, Kały Rybackie, and the Vistula River Museum at Tczew have excellent collections of vessels. Juliusz Struck has a collection of fishing artefacts above his workshop and vessels and old diesel engines outside. For written material – mostly in Polish – see *Polskie Szkutnictwo Ludowe XX Wieku* by Jerzy Litwin (Director of the Polish Maritime Museum) and *Kaszubskie łodzie* by Aleksander Celarek, a sailmaker and boatbuilder from Chalupy, Hel Peninsula. Celarek also wrote an excellent article entitled 'The Barkas of Vistula Bay' in *Classic Boat*, issue no. 35.

CHAPTER 6

The Baltic Coast of Germany

The River Oder to the Flensburg Fjord

The coast is naturally spilt into the three well-defined areas of Holstein, Mecklenburg and Pomerania. Except for the deep-watered fjords such as the Kiel fjord and the shallow Bodden waters, as the inland waters of Rugen Island are called, the wide inshore river deltas were formed following the last ice age. Furthermore the sea, being non-tidal, has a low saline content, which ensures that it often freezes over for two or three months in winter, when making holes in the ice and fishing through them is still common. It's also very shallow coast, shelving very gently in many parts onto sandy, cliff-fringed, beaches. This ensured that the fishermen were largely beach-based, working in the shallow water with shallow-draught vessels and not having to work specifically from rivers with strong tides. Only a small proportion of the fleet was deep-sea based, mostly from villages upon the River Trave at Lübeck and around Rügen Island. Until, that is, the motor arrived!

When one studies the map of the area, the first obvious factor is the proximity of the western portion to the rich Danish fishing grounds. Not surprising, then, that many similarities between the vessels of the two countries can be identified. Indeed,

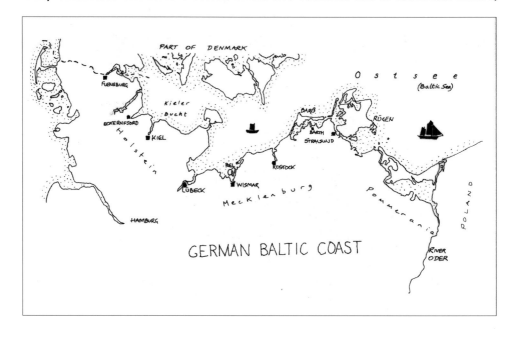

GERMAN BALTIC COAST

Right: Seine-netting was, throughout Europe, one of the foremost fisheries and probably the oldest net fishing method. Here the net has been closed.

Below: The canoes are very simple in construction, though strong and perfect for the job. This is at Gotmund, Lübeck.

Here the boats have come together, the net hauled and are being manhandled to the shore. Note the roller capstans.

Above: A boat at Warthe, close to the Polish border. (*Photo Marina Sundmacher/ Panoramia*)

Left: River craft of the East Prussian waterways.

German half-decked ketch-rigged boats were imported into the Danish Belt region in the 1870s by migrating fishermen. These *zeesenboote*, termed *drivkvase* by the Danes, will be explained in full later.

After about the twelfth century, interchange between the fishing communities of the Scandinavians, Germans, Prussians and Baltic Finns (from Lithuania, Estonia and Finland) led to migration and sharing of experience. These cultural links continue right up to the present day.

The Baltic, as already mentioned, has abundant fish supplies, with herring, cod, flounder, eels and salmon being among the most important catches. Traditional methods of catching were by lining, fixed and drift nets, traps and fish weirs, until the advent of trawling. Fish was either eaten fresh, or dried, salted or smoked for winter use and, surprisingly, these coastal settlements had one of the highest rates of fish consumption in Northern Europe.

A flotilla of craft at a herring seine-net in 1923 at Travemünde.

Boats loaded with herring alongside at Travemünde in 1923.

Finally, before we consider the types, it's worth remembering that, like other parts of Europe, these regional craft were usually named after the port they worked from, or by the type of fishing gear they most often used.

BOAT TYPES

So what of the types of vessel used? Although a large part of this coast was under Soviet control for much of this century, luckily the primary characteristics of Baltic craft have been well documented over the years. A field study by Dr Wolfgang Rudolph, carried out between 1958 and 1965, resulted in his classifying vessels into three groups: log boats, rafts and planked craft – and the latter, where the majority of craft lie, was further sub-divided into three more classifications, which were: bottom shell boats, longitudinally and thwartships-laid bottom planked boats and keel boats, as mentioned in the previous chapter.

Perhaps the best example of a bottom shell boat is the *Rostocker-kahn* ('kahn' = skiff). This boat, generally fished by one man, had two or three planks affixed to the bottom shell. Planks met at each end on a sort of squat conventional stem and stern post affixed to the bottom shell. A fourth plank was added when a wet-well was desired. Rigged with one square-headed spritsail, these craft were numerous in the rivers and estuaries, were up to 6 metres long and had a centreboard.

The second category is again where the majority of inshore craft lie. In the construction of these, the flat bottom is planked up, laying the planks either longitudinally or across the vessel, and then the sides are built up with strakes. Those with transverse planking only account for a few punt-like examples such as the Warnemünde pram, which is a squarish boat not unlike, but larger than, a British duck-punt! However, they were rarely used for fishing. Most were planked longitudinally, such as the *heuer*, the *keitelkahn*, and the Pomeranian beach-based *flounder boat*, also known as the *Dievenowe (Dziwnow) fishing boat* east of the River Oder, across the border in Poland.

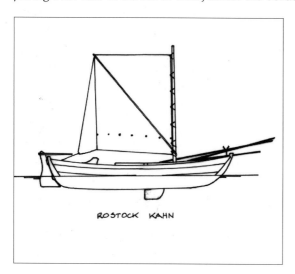

ROSTOCK KAHN

Within this second category, the types can be further broken down through the differing ways of constructing the stem and sterns. The double-ended Pomeranian skiff known as a *Blockpolt* has a block of wood gouged out at either end that is fixed to the bottom planks, and to the plank ends. Similarly, the *Ellerbeck skiff*, a 6-metre box-like vessel with up to two spritsails, has block-like ends affixed with a certain amount of uplift in the sheer so that they resemble, vaguely, salmon cobles when viewed from the side, or *eka* from Sweden's west coast. A fine example of one of these sits in the Altonaer Museum in Hamburg. Likewise, skiffs from Dassower Wiek, Gotmund and Schlutup on the River Trave have stems and sternposts which are 'trunk-like'. These craft, all between 5 and 9 metres, often have centreboards.

Other forms of constructing the stems and sternposts are in a conventional manner, with planked boards with bevelled or square ends.

The Pomeranian *heuer* has one single bottom plank with three or four wide strakes, nailed in a clinker fashion and fixed to conventional posts. No frames were fitted, as was the case in most of these craft, although wooden knees added to the strength between bottom and sides. Thwarts, too, braced the structure. *Heuers* had a wet-well, with holes being drilled in the lowest planks, to keep the catch fresh, and a centreboard. In profile, these craft exhibited a remarkable sheer, and were decked over either end with short decks.

The home waters of the *Schleikahn* are in Holstein. These skiffs, measuring from 23 to 26 feet, exceptionally up to 30 feet, were worked on the River Schlei, and were one of the narrowest of bottom-planked craft.

The Pomeranian *flounder boat*, a flat-bottomed boat with one single gaff sail and foresail, was discussed in the previous chapter. They were generally double-ended, although square-sterned beach boats worked from Rügen. With the mast height being matched to the boat's length, these 10-metre craft were built from oak and pine. In the extreme, some of these craft had three masts. Again, like the majority of these generic small craft, they were used for trapping fish (*reusenboote*) or netting (*garnboote*). Many of those remaining into this century were fitted with motors with the advent of this development.

Other craft within the classification are the *zeeskahn*, again named after the *zeese* net, a sort of trawl net, a *tuckerkahn* and *polt*. Around the mouth of the River Oder, these were flat-bottomed craft, and varied from others by setting lugsails on two masts. Another craft variant of the *zeeskahn* was found in the Rügen-Stralsund area, but, although they are believed to be entirely different, constructional details are scarce. It is known that they were recorded in 1449 and that the last was scrapped in 1935. Herring boats from Rügen were built on a flat keel with one gaff sail and they fished with a particular drift-net called a *Manze*. Smaller beach boats, with five oak planks either side, had two sprits or sometimes gaffs and appeared similar, though much smaller in size, to the Nydam boat, the 75-foot-long oak-built boat excavated near Sonderberg and believed to date from between AD 200 and 400. This type of craft is considered as a precursor to Viking types.

The third category of keeled boats again comes in all sorts of sizes and types. Some small three-masted *jolles*, a type similar to some Danish craft, have survived in some of the fjords and are still used today for pleasure, though the distinct, fine-lined, jolles from the Kiel fjord have entirely disappeared. The open-keeled boats from Warnemünde came in three sizes – the *Volljolle* at 7–8 metres, the *Dreivierteljolle* at 6 metres and the single-masted *Halbjolle* at 4–5 metres. Some replicas of the latter have been built. Similar to these are the *Poeler fischerjolle* from the island of Poel, near

HEUER

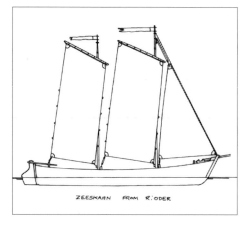

ZEESKAHN   FROM   R.ODER

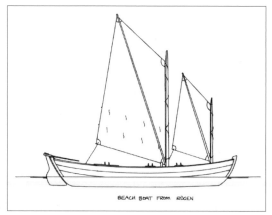

BEACH BOAT FROM RÜGEN

Shore boats of Rügen at Vitt, one of the oldest villages, with its stone harbour wall, which is situated on the north coast of Rügen, in the Baltic. Kap Arkona can be seen clearly in the background. The image dates from around 1920. (*Photo: Foto Knospe*)

Wismar, and these were built on a flat keel, equipped with a wet-well and sprit-rigged like those from Warnemünde. A small number of these, too, have survived and are owned by local 'associations' that have restored them and now sail them on a regular basis. Eckernfjorde became famous for its *jolle* building, with many of its craft being bought by Danish part-time fishermen.

Away from the inshore small craft, several larger keeled boats were traditionally used in the fisheries. Harbours being largely non-existent, especially along the Pomeranian coast, it wasn't until this century that bigger craft were adopted, except for certain examples, as we shall see. However, experiments were tested, with a Swedish *lachsboot* (salmon boat) being tried by the Deutsche Seefischerverein (German Sea Fishermen Association). At the same time a *Blekingseka* from Blekinge, southern Sweden was introduced, as was an English smack to Danzig, but without success.

The *Quatze* was a keel boat which seems to have developed from the *zeeskahn* of the Oder. Not much is known about them, except that the biggest were *seequatze*, which were clinker-built and gaff-rigged. These evolved through being fish-carrying vessels. Rather than the fishermen, when they sailed further into the Baltic, fishing as far off as the Swedish and Danish coasts, having to return home with the catch, the fish was loaded onto the various *quatzen* (plural) with their large wet-wells and returned to market that way. However, fishing boats later had wet-wells fitted so that each boat could stay at sea for a number of days or weeks at a time and subsequently returned with their catch, hence without the need to employ a carrying vessel. The use of the *quatzen* seems to have ended soon after the building of the last vessel just before the First World War, although one remaining *Quatze* has been restored at Rostock. Smaller *haffquatzen* fished in the shallow waters of the Bodden. *Quatzen* were built mostly in Wöllner, Ueckermünde, Neuwarp and Wolgart.

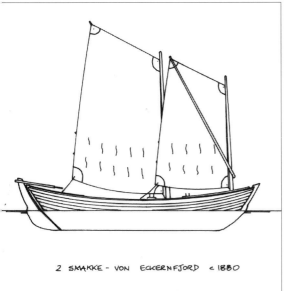

2 SMAKKE - VON ECKERNFJORD c 1880

QUATZE c1910

ZEESENBOOTE

The final type of keeled boat is perhaps the best known and has the highest number of surviving examples and therefore it seems appropriate to give a fuller description. The *zeesboote* (plural: *zeesenboote*) – roughly translating to '*zeese* boat' in English – developed from the aforementioned larger sailing craft, the *zeesenkahn* from the island of Rügen. The first record of a *zeesboote* was in 1858. The *zeese*, the net from which it took its name, was a bag-net that was trawled, originally to catch eels. Like its predecessor, the *zeesenboote* were native to the Bodden, the large area of lakes and inshore waters between Rügen and the towns of Greifswald, Stralsund and Barth. Here they were built locally, first in the fishing settlements and from 1750 onwards in more rural settings. By the beginning of this century they were being built in great numbers in the small village of Freest, where Mr Jarling was responsible for producing some twenty-two such vessels between 1921 and 1946, and in fact remains in business today, albeit building *Ostseekutters*. Another well-known builder was master-builder Dornquast of Fuhlendorf, who built craft for the fishermen in the villages from Barth to Riebnitz, to the west. Several were also built on the island of Poel for fishermen from Wismar who had migrated from Stralsund. It was about the same time that fishermen moved over to the Danish archipelago, as mentioned earlier. There were so many *zeesenboote* fishing in the Stralsund area in 1880 that the city council found it necessary to reduce the numbers from 110 to eighty by introducing bylaws.

Original *zeesenboote* were clinker-built upon a central plank reaching the full length of the keel. This oak plank was about 5 inches high and a foot wide. To this the stem and sternposts, both oak again and curved in the Scandinavian style, were fitted.

ZEESBOAT

ZEESBOOTE
[EASTERN GERMANY]

Later on, a clipper bow and counter stern was adopted. After the rabbet was cut out, the moulds were attached and the bottom four strakes fixed before these moulds were again removed. Further planking continued up to the sheer line, with between ten and thirteen more strakes. These strakes, too, were normally oak. Considerable sheer was produced at the bow, and somewhat reduced at the stern, although the freeboard was kept low for ease of hauling the net. Occasionally fixing of the planks was temporary while the timber dried out, after which they were permanently fixed. Later boats were carvel-built with frames added. Decking and finishing could then proceed. Underneath the foredeck a small cuddy had two berths, a table and stove, while a wet-well was built aft. In the nineteenth century most were fitted with leeboards, until centreboards were introduced after about 1880. This had the added benefit of enabling a larger forecastle to be incorporated.

The usual rig for a *zeesboote* was two pole-masts with a boomless gaff main and standing lug mizzen and fore, jib and topsail. The length of the main was said to be three times the beam of the boat, and masts were often cut and rounded from spruce by the fishermen themselves. The mizzen was set onto the rudder head while

Above left: A *zeesboote* drying its nets, probably at the pier of Stralsund, a town once part of the Hanseatic League of Merchants. The image dates from around 1920. (*Photo: Siftung Pommern*)

Above right: Two *zeesenboote* on the shore of Vilm Island, south-east of Rügen, drying their nets. (*Photo: Siftung Pommern*)

the foresail was attached to a bowsprit. Additional spars (*zeesenbaum*) were fitted at either end of the boat – one alongside the bowsprit and the other reaching long over the rudder – to allow the ropes to the *zeese* net to be attached to their ends, and thus widen the mouth of the net. Once the net was set, always on the starboard side, the boat drifted at right-angles to the wind, and when off the wind, the net could easily be hauled aboard.

On the Dará peninsula some boats did away with the mizzen and were sloop-rigged. Some of these were adapted beach boats, and others bought in second hand. This didn't seem to affect their ability to sail, and all the *zeesenboote* were renowned for their speed. A boat could sail 40 degrees to the wind and achieve 7 knots when running. They could make the journey from Stralsund to Bodstedt in 4 hours, often leaving, it is said, the steam ferry behind!

Regattas were commonplace at the beginning of this century, although, like elsewhere they faded away. In recent years the tradition has been revived by Andreas Schönthier of Althagen, Dará, a fisherman who owns the *zeesboote* 'Sannert', the last one that fished. Along with Eckehardt Rammin of Bodstedt, an annual event is organised in each September, and more and more of these fine craft have been lovingly restored so that numbers increased as did interest in their cultural roots. Today the remaining *zeesenboote* are recognised in their own class as members of the Bund Deutscher Segler der DDR (although now the DDR bit has disappeared) and they all have their FZ registration number. In 1980 there were twenty-eight boats, and the number has grown steadily ever since and now stands at about sixty and they are now an object of cultural interest. Eckehardt Rammin now has his own yard at Barth where boats are maintained, restored and rebuilt. Throughout the summer months the *zeesenboote* are in action in various ports, so that they remain one of the most prolific in terms of surviving numbers of yesterday's sailing fishing boats today.

Two clinker transom-sterned boats used to service set nets around the inland waters of the former East Germany.

MOTORISATION

Motorisation affected the fishing fleets here much like the rest of Europe. The *wadboat* of Rügen, an eel boat, became one of the first to be engined, during the First World War. These evolved quickly into the *Rügenwalder fischkutter*, which show a strong resemblance to the Swedish 'koster' boats. The first *zeesenboote* to be engined was in 1927. At the extreme western end of this coast, the *Maasholmer fischkutter* appears similarly Scandinavian. At the same time many beach boats had motors installed. All these boats retained their rig for a time after motorisation. Further along the coast in the Lübecker Bucht the *Ringwadenmotorboot* worked a ring-net and these, uncannily, show a strong similarity to the ringers of the Clyde, in western Scotland, and were unrigged. Even under the Eastern German Socialist Government, then, the ubiquitous motor boat had arrived for good here as elsewhere!

The motorised fleet of cutter-type craft at Eckernförde in the early 1950s.

SOURCES

The Nautineum Museum of the Deutsches Meeresmuseum in Dänholm, Stralsund, has an excellent collection of historic vessels, half of which are housed in a recently-built enclosed building. Others are to be found in the grounds, as are examples of fishing nets. Additional information comes from: *Bootsbau* by A. Brix (Hamburg, 1990); *Zeesenboote im National Park* by A. Dietzel, E-U Krohn and R. Legrand; *Zeesboote* by Hermann Winkler, *Fischerei in Schleswig-Holstein* by Horst Shubeler, *Segelboote – der Deutschen Oosteekuste* by Wolfgang Rudolph (which has an excellent bibliography) and the National Maritime Museum, London, Monograph No. 14, as mentioned in the sources of a previous chapter.

Two boats at the Nautineum Museum of the Deutsches Meeresmuseum in Dänholm, Stralsund, one a canoe type, the other a beach boat.

CHAPTER 7

Denmark and Iceland

*The Flensburg Fjord to the Island of Sylt
and Iceland and the Faroes*

Because the country of Denmark is a peninsula surrounded on three sides by water and a jumble of some 400 islands (only ninety-seven of which are inhabited), it has, relative to its small size and population, a long coast line of nearly 4,600 miles. On one side is the fish-rich North Sea and on the other the quieter, yet still rich, waters of the Baltic, where the islands are clustered around its southern part. Denmark acts as a barrier between the two seas, allowing narrow seaways for shipping and sea-life to enter and leave. These waters, the Skagerrak and Kattegat, are themselves also full of the rich resource of fish. The channels between the islands are known as the Belts, while the narrow seaway between the large island of Sjælland and the mainland of Sweden is the *Øresund*, literally 'The Sound'.

The country has had, over many generations, an abundance of small craft that have evolved over generations from when the warrior-like Danes aboard their longships were feared visitors to the neighbouring coastal settlements of northern Europe. However, from a fishing point of view, these designs have developed through a local need. Norwegian vessels obviously influenced areas, due to the shared fishing grounds in the Skagerrak and beyond. Most boats were built locally, with the timber coming from nearby forests and with the fishermen often lending a hand in the process to save money, especially in the sawing up of the timber baulks, the rigging and the painting. Up until the twentieth century, boats were clinker-built of pine, oak and larch.

Denmark has had a good number of ancient boat finds excavated over more than a century, representing both pre- and post-Viking craft. The majority of these are large sea-going craft, though one smaller vessel found in Roskilde fjord, where the five 'Skuldelev boats' were excavated in the 1950s, is thought to perhaps been a fishing vessel dating to about AD 1000. The earliest boat found and partly excavated, and already mentioned in the previous chapter, is the Nydam boat of AD 200–400.

THE BALTIC

Herring, as already described, was plentiful in the Baltic in the fourteenth century, whereupon the Hanseatic League of North German Merchants was formed. Over the next few centuries thousands of boats fished the herring, centred on what is now the south-west of Sweden, which was in fact part of Denmark at the time. When this dominance of the fishery was broken in the late sixteenth century, the herring became the despised diet of the common folk. Peasant farmers called it their 'herring breakfast'. The Danish fishermen became poorer and fewer up to 1800, when their fortunes altered for the better. With the dogged perseverance for which fishermen are renowned, good catches of herring were taken off the Norwegian coast in the early part of the century, and later in the Skagerrak. The design of the herring boat had remained unchanged for ages. The basic *sildebad* – herring boat – was a light, keel-built, double-ender with an S-shaped midship section. Herring was caught at night in drift-nets during the spring and autumn herring seasons. This basic boat shape remained the same right up to motorisation, though they also participated in other forms of fishing and even some cargo-carrying work.

In the Øresund, where the herring fishery was centred at Skovshoved, a typical boat was 7 metres in length and had a narrow form, ideal for the short seas there. They were sprit-rigged, which accounted for their speed. However, on the north coast of Sjælland, at Hornbaek and Gilleleje, larger, much more beamy craft with a gaff rig were used in the more open waters of the Kattegat. Like the Norwegian types, these had a cutaway forefoot and heel, long keels and a full form above the waterline to match their beam. They were sharp below the water to ensure speed. Similar vessels still work from Hornbaek though instead of sails, these have been equipped with motors.

In the nineteenth century many Øresund fishers worked in the waters of south Sjælland and Funen, in an area known as *Smalandshavet*. The *Storebelt* – the Great

A two-masted sprit-rigged boat from around the Great Belt between the islands of Funen and Zealand. (*Photo: Danish Fisheries Museum*)

Belt – fishermen were attracted by these trim, seaworthy boats and so adapted their *sundebad* – sound boats. These sound boats were either open gaff-rigged 7-metre-long double-enders or transom-sterned boats of a similar length, the latter being regarded as the older of the two types. Many new vessels were built, modelled on the Oresund boats. At the same time other boats were brought in from the north of Sjælland. These narrow-beamed *belt-jolles* drifted with twenty to forty herring nets. Many were later adapted with wet-wells for flat-fishing.

BORNHOLM BOATS

In the small island of Bornholm, to the east of the southern tip of Sweden, fishing has always been an important part of the local economy and has, over the last 500 years, been controlled by various powers such as the Danes, then the Lubeck merchants, the Swedish Crown and then Denmark again. Thus various influences have come together to produce a varied boat heritage. The rocky, rugged coast is quite different here to the rest of Denmark and the earliest documented boat was the *ege*, or 'oak' boat, which were commonest up to the late nineteenth century. The curious rig had a dipping lug main with mizzen and fore sprits, the latter being set on a short mast raking forward at an extreme angle. They were transom-sterned, clinker-built and ranged from 7 to 9 metres in length.

Another type peculiar to Bornholm was the small, open double-ender with a single sprit, two foresails and, sometimes, a topsail. These *saettebad* were used for the herring in late summer and the cod in winter. Their name, which literally translates to 'putting-out boat', comes from their habit of setting the net one day and hauling it in the next. They remained in use throughout the nineteenth century.

The first decked boats in Bornholm were built there in 1867 and were inspired by the government fisheries consultant and actor A. J. Smidth, and by drawings by the renowned shipwright E. C. Benzon. They were generally between 10 and 12 metres and were chiefly used for catching salmon and cod during winter. Regarded as seaworthy vessels, some sailed as far as Iceland and often into the North Sea. However, they were regarded as being too heavy for the summer herring. After a few decades, smaller half-deckers of 7–8 metres superseded them and became all-year-round boats.

These were gaff-rigged, with topsail, staysail and jib, and they chased the herring in the *Smalandshavet* and the west part of the Baltic alongside the boats of the Great Belt, following the autumnal migration of the herring. The Bornholm boats, because of their better qualities, were bought by the locals and amalgamated into their fleets until local boatbuilders, especially those from Lohals, began to produce their own versions. This was after 1900, and so coincided with the beginnings of the motorisation of the fleets. Hence these new boats – often referred to as the *Lohal boats* – simply reduced the rig to a gaff main and job and had, typically, a Dan 15 hp motor fitted.

At the same time as motorisation was changing the face of fishing, a more modest and cheaper version of the Lohal boat was built and became known as the so-called *Nyborg boat*. Together, these two types, because of their motors and greater tonnage, commanded the fishing as the earlier herring boats were quickly withdrawn and phased out, and this command lasted right through to the 1960s. Not only that, but their use spread into the Øresund, replacing the smaller herring *jolles*, and together with the motorised Skovshoved boats, they were largely the only types of boats used for fishing.

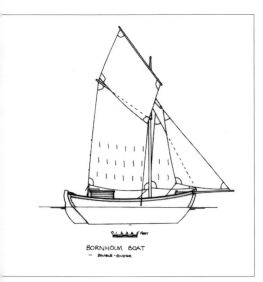

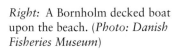

BORNHOLM BOAT
– DOUBLE-ENDER

Right: A Bornholm decked boat upon the beach. (*Photo: Danish Fisheries Museum*)

SOUTHERN FJORD BOATS

Small *jolles* also worked from the southern Danish waters. These *smakkejolles* – two or three-masted yawls – were small, open, sharp-sterned craft. They were used for small-scale herring and mackerel fishing in the *Lillebelt* – the Little Belt – and for ferrying work and pilotage among the islands. Although they were small, say 6 metres long, they were able to carry three masts when working in the sheltered waters of the area, though one mast could easily be lowered in the event of the wind suddenly increasing in strength. Many were built in the Eckernförde, in Germany, as discussed in the previous chapter, as well as in Danish yards around the Great Belt. They carried very square-headed sprits in the nineteenth century, though these were 'peaked up' in the twentieth century. For fishing purposes their use declined between the two World Wars, when decked vessels began using trawls and stow-nets.

Flat-bottomed boats worked some of the inland parts. These were almost identical to the German *kahn*, described in the previous chapter. On the Isefjord, on the north coast of Sjælland, eel fishing was undertaken in *prams* while a seaworthy small *jolle*, principally from the small town of Lynaes, was used for fishing, pilotage, life-saving and serving nearby lighthouses.

In Denmark, drifters were known as *drinkvase*. These flat-bottomed boats were used to seine-net for eels at night around the southern parts of the country. They drifted sideways with the net suspended from the bowsprit, sails aback with a small mizzen set to balance the drift. In the 1870s, however, German fishermen from the Pomeranian coast – especially from Rügen and Stralsund – moved to this coast, bringing with them their ketch-rigged, counter-sterned gaffers. Local boatbuilders then began to copy these *zeesenboote*, combining the German hull with the Danish profile to produce another drifter. These were deemed suitable for motorisation, and their use continued until recent times.

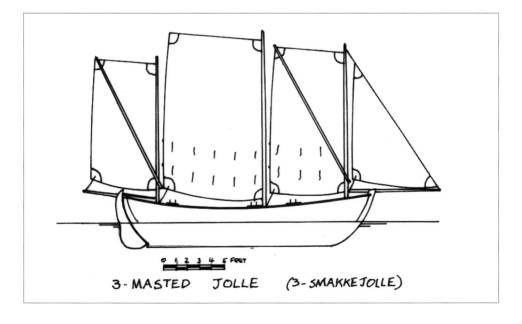

3-MASTED JOLLE (3-SMAKKEJOLLE)

THE NORTH SEA COAST

The west coast facing the North Sea is flat and sandy, with sand dunes backing a largely barren hinterland. It's an unprotected shore, with the sea constantly buffeting it. Fishing, here, was a part-time occupation. Smallholders and agricultural labourers would come to the coast in spring and reside in huts built near the shore until the end of the fishing season. In a typical camp three huts housed the six or eight fishermen, the six bait girls who prepared the lines, two bait fishers and one runner who sold the catch. Another hut was kept for the fishing gear. The bulk of the fish was sold locally unless they were lucky enough to spot a passing trading smack, which would often anchor offshore and buy up the entire catch. When the railways arrived in the late nineteenth century all this changed and much of the fish was sent directly to markets in northern Germany.

Up to this point the small beach boats had remained largely unchanged for generations. The oldest of them was the *havbad* – meaning 'sea boat' – which had survived from very early times. They were 7–10 metres long, with heavily raked stems and sternposts, and had a small keel. Normally they were rowed by four, six, or even eight oarsmen. A small spritsail and jib were added at some time before the late nineteenth century, a time when the craft were enlarged. They were mostly used for long-lining for cod and haddock.

Hjerting, the main fishing station to the south before Esbjerg's harbour was built in 1868, had its own *pram*-type, flat-bottomed boat, a so-called dory, under 6 metres, usually with four planks either side. Broad and stout *prams* had been introduced from Canada and these were distinguishable by their flat bows and sterns, although the Danes adapted the type to suit themselves. When Hjerting's prams had lugsails added in the nineteenth century, a centreboard counteracted the rig and, with a 1.7-metre beam, these fine-lined craft were very fast.

Following the construction of Esbjerg's harbour, Hjerting's fishermen based themselves there, and instantly adopted a larger, decked boat with proper sea-going qualities, suited to a harbour and not a beach. These had a central open hatchway and a cutter rig in favour of the sprit, with a boomed main. The shape of these double-enders proved so popular that even larger boats were imported from Norway in the late 1870s, until local boatbuilder Th. Dahl began to build his own version. Four hundred were eventually built for the line fishery by 1879. Use of these decked boats had spread rapidly along the entire western coast of Jutland by the end of that century.

Further north, at Vorupør, another sprit-rigged *pram* remained more like the Norwegian types. These transom-sterned boats, with an uplifted bow and round hull, were about 3.5 metres long overall. Similar *prams* of 3.5–5 metres, with semi-circular sterns, worked off the northern tip of the country, lining and lobstering. Many flew a topsail as well as spritsail and jib.

Klitmøller lies a few miles north of Vorupør, and the *jolle* of that name had a transom stern with a small keel. The rig was like that of the *Vorupør pram*, but otherwise these boats were peculiar to Klitmøller. They remained in use up to the 1960s, albeit altered after motorisation.

The typical sprit-rigged *havbad*, meaning 'sea boat', the most common boat working off the North Sea beaches. (*Photo: Danish Fisheries Museum*)

West coast *prams* from Vorupør, around the turn of the twentieth century. (*Photo: Danish Fisheries Museum*)

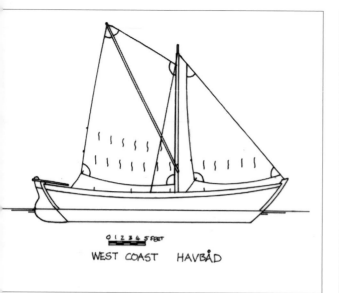

WEST COAST   HAVBÅD

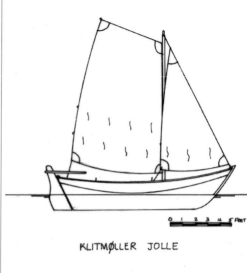

KLITMØLLER JOLLE

92 A sprit-rigged Klitmøller *jolle*. Note the man with a fisherman's model. (*Photo: Danish Fisheries Museum*)

Jolles on the beach at Klitmøller in the north of Denmark in the early twentieth century. (*Photo: Danish Fisheries Museum*)

The Limfjord divides the northern part of Denmark from its much larger southern mass, and is rich in seafood. In these shallow waters the *kag*, a flat-bottomed vessel, was ideal in the sheltered eastern waters, where they fished with hooks and lines. To the west, where the fjord opens up, the *kags* had narrower flat bottoms for better stability, and they set a spritsail and jib. Western *kags* were 8 metres in length while the eastern versions were only some 6 metres.

However, fishing was banned in the fjord during a fair portion of the year so that the *sjægt*, or skiff, was introduced to combat this prohibition. Designed and built specifically as a fast boat to evade the detection of the fishery officers, these half-deckers were 6 metres long, sprit-rigged, and had a cut-away forefoot and keels in the Norwegian fashion. A classic evolutionary process occurred here after motorisation. Motorboats were confined to being under 5 gross tons to prevent the larger boats from plundering the fjord, and so local fisherman Anthon Jenson, who was from the main fishing centre of Glyngore, had built in 1926 the first of what was to be called a *pennalhus*, which translates as 'pencil case', on account of the elongated, straight shape. This first boat was *Proven*, A721, which was 8 metres long by 2.5 metres in the beam. She had an almost vertical stem, a shallow forefoot, a wet-well, a removable wheelhouse and a 10–12 hp Grenaa petrol engine. The other fishermen soon adopted the design, and the boats became popular throughout the fjord.

The Danish seine-net was also invented here in 1848, a mode of fishing that impacted on the fishing fleets all over the North Sea. This is generally recognised as being invented by fisherman Jens Laursen Vaever from Salling in Jutland. Like most of the farmer/ fishermen from the Limfjord, after the sea broke through the western sea defences in 1825, allowing cod and plaice to invade, Vaever caught these fish using a *Kratvoddet*,

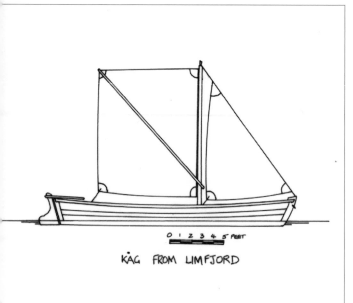

KÅG FROM LIMFJORD

SJÆGTE

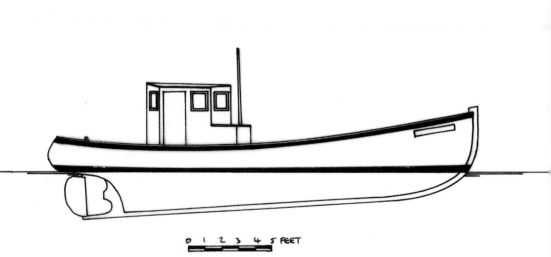

'PENNALHUS' FROM LIMFJORD - 1926

a large beach seine net run out from the shore using a small rowing boat and fixed by one end to a post. In the same way that the herring fishermen of Tarbert, Loch Fyne, experimented with beach seines to develop the ring-net in the 1830s, Vaever, in the late 1840s, experimented using a net in the same shape as the Kratvoddet, running it out as usual but then hauling the net back to another anchored boat. His first attempt was disastrous, much to the bemusement of the on-lookers, although that changed to chagrin on the second attempt when Vaever landed 2,640 plaice. The fishermen working from the beach were happy to land forty or fifty mainly plaice a day. When Vaever landed 4,000 on his third attempt, there was a rush to follow his example. It is said that over his first two days of fishing he earned enough money for his wedding to Anna Marie Neilsdatter. Vaever was one of the great innovators in the fishing industry of the last two hundred years, and in an industry slow to change, this was relatively rare.

Vaever's invention, which became known as the *snurrevod* (*snurre* being the rotating action around the anchor and boat and *vod* meaning net), spread to the east coast by 1870, to the northern fishing stations eight years later, Esbjerg by 1890, to the Bohusland coast of Sweden in 1894 and to Britain in 1920. This development was largely responsible for the 1880s boom in the Danish fisheries. Inshore fishing extended itself into deeper water and a fish manufacturing base grew up quickly about northern Denmark.

In Frederikshavn, originally the small fishing village of Fladstrand, which gained a municipal charter in 1818 and changed its name and was the major fishing harbour for both the Kattegat and Skagerrak, the fishermen were among the first to adopt the new method. However, as the offshore fishing developed, it was soon clear that it was possible to fish the *snurrevod* in the open sea, for which they brought in large so-called cutters. By the late 1870s these were ketch-rigged, fully decked vessels of 40–60 tons, some of which had been bought from England's east coast at a time when the first steam trawlers were being introduced there. Others were built locally, modelled on similar lines, when the first shipwright called Buhl had commenced building. Many of these new vessels had wet-wells incorporated into their hulls. Such a vessel was the *N.I.Laursen*, FN136, rigged as a gaff ketch.

Of course, it didn't take long for the other principal ports of Denmark to adopt the seine-net, which became known as anchor seining for obvious reasons. Esbjerg, which had its first harbour works started in 1869, had similar-sized cutters working from there by 1888 and within six years the method had spread to the west coast of Sweden. Larger boats used a small 20-foot seine boat – a *snurrevodjolle* – with which to work the net out and around in a circle before it was brought back to the cutter. A typical seine boat was modelled on traditional double-ended small boat lines and around 1903 the first engine to be fitted into a fishing boat was installed in one of these. The first engine was a Mollerup 2 hp unit and the long term effects of the petrol/paraffin internal combustion engine arrived. It didn't take long to catch on.

The Frederikshavn men built smaller cutters, oak on oak, with motors which obviated the need to use the *snurrevodjolle* to set the net, the first carvel-built example of which was the cutter *Gorm*, substantially smaller at just under 10 metres in overall length. A smaller rig was retained, and these boats had greater effectiveness and were easier to manoeuvre

SHARK CUTTER c.1900

and became known as the *haj-cutters*, literally 'shark cutters'. Most of these retained the wet-well and carried ice from the new ice houses. Shark cutters arrived in Esbjerg in 1910. However, some of the larger 50-foot cutters, generally referred to as *kotters* today by many, continued being built, especially around Frederikshavn and other parts of northern Denmark, some of which have survived today and have been put back to a full sailing rig.

After the First World War, the Danes began fishing the western side of the North Sea and consequently landed their catches into English ports. Grimsby men observed their gear and began using the same method on their steam trawlers. In Scotland, the first time the fishermen used the method was in 1921 and six years later the first Scottish seine-net vessel built specifically for the seine net, the '*Marigold*', was launched on fifie lines from the yard of William Wood & Sons in Lossiemouth.

The last development in the Danish anchor seiner was the adoption of the cruiser stern, which appeared some time before the outbreak of war in 1939 and the subsequent invasion of the country by the Nazis. Although the canoe stern and variations of the cruiser stern appeared in the 1920s, the anchor seiners appear to have been content with their counter sterns until, presumably, engine power increased and the cruiser stern was deemed a better option. Thus the typical Danish seiner arrived, a boat that impressed the Grimsby men so much that they, too, adopted similar vessels and called them 'snibbies'.

Boatbuilders and fishermen tell stories of visiting Denmark to buy up decommissioned kotters in the 1970s and 80s, when hundreds of these vessels were withdrawn from fishing. Some talk of boats being offered for sale, both counter and cruiser-sterned, on the basis of 'buy one, get one free'! These boats, built solidly of oak with massive scantlings, were perfect for conversion to pleasure and charter boats and many still sail. Thus many examples of these kotters remain in different parts of Europe. '*Mias*' was built at the Bronsodde Shipyard in Vejle, Denmark in 1934 and registered as RI100. She fished out of her home port of Hvide Sands under her owner, Jens Viggo Iversen. When she was sold to Niels Christian Andersen in 1946, he changed her name to '*Inge*'; the following year she was renamed again to '*Inge Wejse*'; again to '*Anna Katherine*' in 1962; and to '*Vilma*' in 1964. In 1979 she was sold to William Rarity of Sunderland, who changed her name to '*Wilma*' and registered her as SD112. She was sold yet again in 1987 to John Watson of Maryport, and the name changed to '*Cee Kay*' and registered as LL36. Today she is named '*Vilma*' and sails as a topsail schooner out of Bangor, North Wales.

The 1931-built '*Josefine*' was launched as the '*Lilly*', L60 from the Andersen & Ferdinandsen yard at Gilleleje, and nowadays charters around the Bristol Channel and beyond. '*Tangaroa*', built as '*Tina*' by Soren Larsen, Nykobing Mores in 1925, has recently moved to Bristol, where she is chartering. The '*Eda Frandsen*', built in

The shark cutter fleet leaves Esbjerg in the winter of 1910. (*Photo: Danish Fisheries Museum*)

A typical shark cutter remains one of the prettiest of the early motorised vessels in Europe. (*Photo: Danish Fisheries Museum*)

Above: The shark cutters were excellent sea boats, many fishing out of British North Sea ports. During the war many crossed the North Sea to take refugees from Nazi-occupied Denmark.

Left: The 1934-built *Vilma*, originally *Mias*, rigged as a Welsh topsail schooner by her owner boatbuilder Scott Metcalfe of Port Penrhyn, Bangor, North Wales, firing a salvo at the 'Gaffers at Holyhead' maritime festival in 2009.

Grenaa as '*Lagoda*' in 1938, works off the west coast of Scotland. '*Lola of Skagen*', built by the Nipper Shipyard, Skagen and launched in 1919, works along the west coast of France. Others include '*Genara*', built at Frederikshavn in 1926 for Grimsby and registered as GY382; '*Ronsus*', possibly built by Otto Dolmer, Lesoe in 1910; '*Grietje*', built in 1932; '*Annette*', built in Skagen in 1908; and finally '*Ceylon*', WA18, built at the Nipper Shipyard, Skagen in 1939 and originally fished from Thyboren as '*Ceylon*', L246. She was later sold to Lysekil, Sweden, where she fished as '*Ceylon*', LL252 until being sold to Whitehaven around 1970/71. '*Annette*', her original name unclear, fished until the 1970s, at which time she ended up in Jegindo, an island in the Limfjord, where she was bought by local boatbuilder Jan, along with another kotter, and he converted both vessels, renaming her '*Annette*'. She arrived in Britain in 1998 and is currently in Gweek. The '*Andreas Jensen*' was built by the Karstensen & Henriksen Shipyard in Skagen in 1945 and first registered as S320 under the ownership of H. F. Jensen. The '*Corona*' was built at the Jensen & Lauridsen Shipyard in Esbjerg in 1931 for S. Christensen and registered as E468. The boat stayed in his family for 25 years before being sold in 1956. By 1970 she was owned by E. S. Pedersen and had been renamed '*Bente Stenberg*', E468, while Vagn Pedersen (possibly a relative) had bought her by 1977 and renamed her again as '*Minna*', E468. Three years later she was under British ownership, working as '*Leason*', GY440, out of Grimsby. She eventually moved around the country and is now in Bristol under restoration. Several still work in Swedish waters, such as the 1093-built '*Hvitfeldt*' or the 1940-built '*Duen*'. Likewise, the 1936-built Norwegian '*Nordhav*' displays similar lines, for these boats were built in both Norway and Sweden, although not in the same number as in Denmark. Furthermore, the 1931-built, red-painted '*Dagmar Aaen*' originally hailed from the N. P. Jensen yard of Esbjerg and fished until 1977. She was then bought and restored and has since sailed over 150,000 miles, much in the polar regions. Her hull, even though the oak frames are spaced only inches apart, has been strengthened with aluminium and steel for sailing within the ice. She continues sailing today under the ownership of explorer Arved Fuchs, who runs expeditions into the Northern polar region each year. The vessel, so I'm told, is the only boat to ever navigate the North West passage without ice-breaker support. Many examples of the later cruiser-sterned boats still fish.

ICELAND AND THE FAROES

Denmark has had considerable influence over both Iceland and the Faroe Islands. Iceland was part of the Norwegian kingdom, though suffered economic and cultural setbacks in 1262 because of disputes between Denmark and Norway. It was a supplier of fish to Norway until 1397, when that country, as well as Sweden and Denmark, became part of the Kalmar Union under Danish dominance. Denmark did not have the need for its fisheries and, to aggravate the situation further, between 1602 and 1854 Iceland was prohibited from trading with any other country except the homeland. Thus it was an undeveloped country by European standards, with fishing

Left: Deck view of the 1931-built *Leason*, originally *Corona*. She has recently had a new keel fitted as part of her restoration in Bristol.

Below: The cruiser-sterned boat *Lola* during the Second World War, still fishing.

similarly primitive. With the country bereft of trees, boatbuilding was largely non-existent. Once political control was granted in the late nineteenth century, and with full independence in the twentieth, the fisheries soon developed.

The boats the fishermen did use were mostly imported from Norway and Denmark. By the sixteenth century these were rigged with one square sail, and in the nineteenth century the spritsail was adopted. Boats had an extreme, cut-away forefoot and curved sternpost, and were about 7 metres long. The use of oars persisted because of the fickle winds in the fjords. Seldom did they venture into the ocean. In 1786 a landowner, Skuli Magnusson, attempted to improve the economic life by buying in Norwegian craft. These were open boats with anything from two to twelve oarsmen. During the first decade of the nineteenth century some fishermen used decked-over sail boats built in Iceland for whaling. In 1828 there were sixteen such vessels of 8–15 tons, and this had risen to twenty-five by 1853. Two-masted schooners must have proven more profitable as there were sixty-three of these, all from Denmark. Then, in 1890, ninety English smacks were brought in.

Small boats, however, remained in use in the bays of the west coast. These boats were 5–7 metres long, had two masts and either spritsails or gaff mains and mizzen sprits. They set a staysail and a jib on a bowsprit, and were basically the old square-sailed boats brought up to date and called *sketa*. One such boat had a Mollerup motor installed in 1906, and by 1930 there were 600 motor boats.

The first steam trawlers working in Icelandic waters were British, starting in 1891. Then in 1905, the first Icelandic-owned steamer arrived from Britain. By the following year six were registered, and then twenty by 1915. The subsequent story of Iceland's fishing is well known: plundering by British trawlers and others, culminating in the Cod Wars of the 1970s, which resulted in Iceland gaining full control of her 200 mile territorial waters.

In the Faroes, the fishermen retained their old clinker rowing boats, 20–33 feet long, called *attamannafar*, or 'eight man boats', up to the end of the nineteenth century. Formerly these were rigged with a single square sail, until British influence led to the

ICELANDIC SIX-OARED
OPEN BOAT

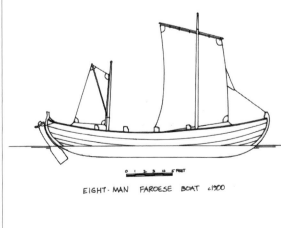

EIGHT-MAN FAROESE BOAT c1900

Boatbuilder Johan Danielsen, who constructed about 1,000 boats in the Faroes. (*Photo: Faroese Museum*)

dipping lug and mizzen sprit. In the 1870s English smacks were introduced, and by 1895 there were 126 Faroese sloops, as they became called, with an average of 80 registered tons. These were chiefly used for fishing for cod in Icelandic waters.

The first steam trawler was purchased in 1922, but three-masted gaff schooners, like those used in Iceland, proved more effective, so that by 1929 there were seven such vessels. That year the Faroese fleet consisted of these schooners, one steamer, 158 Danish 'kotters', 144 motor boats and some 1,500 open boats. Surprisingly, the open, non-motorised boats survived in use well into that century.

SOURCES

The two classic books on Danish traditional boats are *Wooden Boat Designs* by Christian Nielsen and C. F. Drechsel's *Oversigt over vore Saltvandsfiskerier*, the latter first published in 1890 and also explaining the fishing methods. The National Fisheries Museum at Esbjerg has a wealth of information and has published various books, including *A Century of Danish Fishing* and a reprint of the above-mentioned book by C. F. Drechsel. There's also a fishing museum at Grenaa. The only information I came across concerning the Icelandic fisheries was through the Fishing Cultural Heritage Network website 'Fishnet', which can be found through www.svs.is and has some newsletters and papers available online. There is, however, a museum dedicated to the herring at Siglufordur and maritime museums at Rekjavik and Eyrarbakki.

PART TWO
THE ATLANTIC SEABOARD

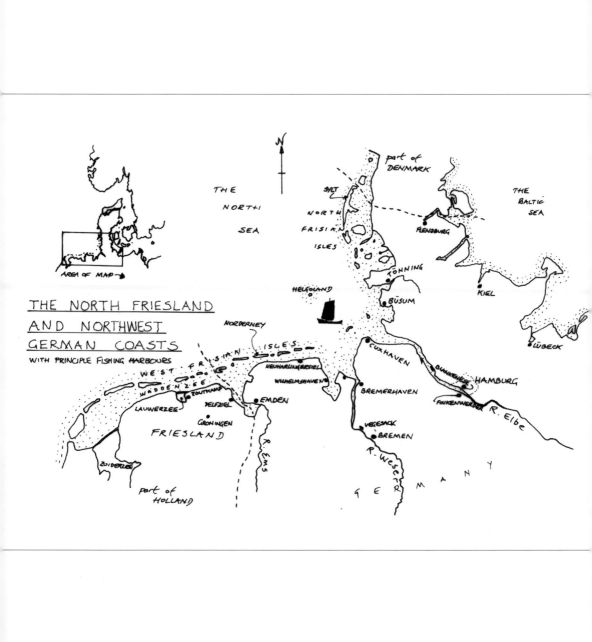

THE NORTH FRIESLAND
AND NORTHWEST
GERMAN COASTS

WITH PRINCIPLE FISHING HARBOURS

N

AREA OF MAP →

THE
NORTH
SEA

THE
BALTIC
SEA

part of
DENMARK

SYLT

NORTH
FRISIAN
ISLES

FLENSBURG

HELGOLAND

TÖNNING

BÜSUM

KIEL

LÜBECK

NORDERNEY

CUXHAVEN

WEST FRISIAN ISLES

NEUHARLINGERSIEL

WILHELMSHAVEN

BLANKENESE HAMBURG

WADDEN ZEE

ZOUTKAMP

DELFZIEL

EMDEN

BREMERHAVEN

FINKENWERDER R. Elbe

LAUWERZEE

GRONINGEN

VEGESACK

BREMEN

FRIESLAND

R. EMS

R. Weser

ZUIDERZEE

part of
HOLLAND

G E R M A N Y

CHAPTER 8

The North Sea Coast of Germany

Sylt to the River Ems

Both the geographical and socio-cultural characteristics of the North Sea coast of Germany differ considerably from their Baltic counterpart in a variety of ways. Briefly, we find here that the Romans some two thousand years spoke of the 'Frisli' as being a tribe 'living here in terps along the coast' – 'terps' referring to their particular abodes. In the Middle Ages the Friesland territory stretched all the way along this coast of the North Sea, from Zeeland right up to present-day Denmark, though by the sixteenth century this had shrunk to about its size of today. Little surprise, then, that Dutch boatbuilding techniques have influenced the design of fishing craft over the generations hereabout.

Shrimp fishing has for long been an important part of everyday life between the Waddenzee and Lauwerszee (the sea between the islands and mainland and the inland estuary of the River Reitdiep respectively) – that is, before the latter was closed to the sea at Lauwersoog in much the same way as the Zuiderzee was closed off by a manmade dyke. Zoutkamp was the main port, with much of the catch being sent direct into the interior of Germany, though it is now inside the Lauwerszee. With the loss of this market during the First World War, small shrimps were dried and sold locally as duck food, such was the surplus. A bit like using herring as manure during times of plenty, as often happened in many parts of herring-rich countries at one time.

The herring fishery did flourish here in the early nineteenth century, with much of the catch being taken to Groningen. However, the smaller boats were unable to compete with the Scheveningen and Zuiderzee *loggers* after the repeal of the Dutch Monopoly Act of 1857, so that after about 1870 these fishermen were crewing aboard both Dutch and German large herring boats.

Meanwhile, the inshore fishermen worked their *blazers*, which will be described in the next chapter. The Moddergat fishermen, for example, caught cod, haddock and flatfish with their *hoekwant* – long, heavy lines baited with lugworm set on numerous cross-lines. When the entire Moddergat fleet was sunk in a gale in 1883, the centre of this fishing moved over to nearby Delfzil, where it survived well into the twentieth century.

The German coast hereabouts is low-lying, with shallow, sandy water and fast tides between the chains of islands and the coast. All avid readers of Erskine Childers' masterpiece *The Riddle of the Sands* will already know all about these hazards for it was here that his two intrepid sailor-heroes, when faced with the might of the developing German navy, dashed in desperation across sandbanks in gales and rowed in fog over the watersheds, followed by spies aboard the *galliots* and eventually unearthed a dastardly plan to invade Great Britain. What a cosy, snug little boat '*Dulcibella*' must have been!

Harbours here are mostly man-made and the whole coastline is protected by a dyke that retains the higher sea from flooding the low-lying, largely reclaimed area at high water. The harbours are either built as 'siels' – gates that open at low water to allow the escape of drainage water, but then close when the tide rises to prevent flooding – or 'wardens', which are man-made hills, some 2 or 3 metres above the highwater mark, upon which buildings and a harbour can be built. The latter are generally found to the east of Wilmhelmshaven, the former to the west, and hence the towns have names such as Neuharlingersiel and Langwarden.

One specific type of craft worked these shallow waters. The *schaluppe* was primarily a Friesland boat and was bluff and squat like the Dutch craft, though a bit smaller than the archetypal Dutch *bomschuit* (which we will learn about in the next chapter) at 12 metres by 5 metres in the beam. They had round sterns, leeboards and a single spritsail. Specific variants were the *Norderneyer schaluppe* – roughly translating to 'Norderneyer line boat' and which were later adapted to gaff-rig – and the *Angelschaluppe* of Helgoland – literally 'fishing sloop' but still referring to line or lobster-potting boats. Both these were used for trade as well as fishing, the latter hardly being over 9 metres in length while those engaged in trading might be twice that length. The *Angelschaluppe* had the distinctive tradition of having a painting on the bow. Rigged with a single spritsail, one characteristic of this rig was that the mast raked forward. Forward of this a foresail was set on a long bowsprit, and leeboards were retained to counteract the rig. Today in Helgoland, the *hummerboot* is reminiscent of the older sloops, and they are still used in the age-old traditional practise of being exclusively allowed to ferry passengers ashore from the numerous ferryboats from the mainland. Being open, clinker-built craft, they are only used close to the coast, sometimes to fish, and are generally fitted with motors. But like elsewhere, the ubiquitous motorboat surrounds those few *hummerboot* that do still remain.

The main fishing communities along this coast, though, were centred upon the three great rivers that drain this part of Germany and its hinterland – the Ems, the Weser and the mighty Elbe. On the lower parts of the latter two, the *dielenboot* seems to have been the fishing boat most in use. Unable to translate properly, it seems that the nearest literal meaning is 'floorboard', and I wonder whether this refers to it being a board boat as against a logboat, or one that was often built using the remnants of building materials – especially the flooring! More likely is it being a colloquialism.

However, they were open boats of about 5–6 metres with a characteristic pram-like stem with a forecastle and a transom stern, and were fitted with leeboards. Some were two-masted with lugsails and a few had three masts with the addition of a mizzen sprit. Larger versions were called *dielenschiffs*, which were about 8 metres long. After the First World War they were all motorised and of a similar size, retaining the rig but losing, to some extent, the pram bow. Mackerel boats, on the other hand, were of a similar size but rigged with two spritsails and a jib.

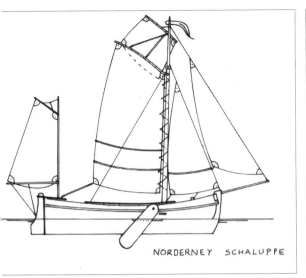

NORDERNEY SCHALUPPE

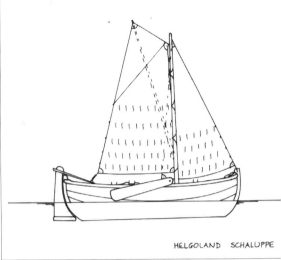

HELGOLAND SCHALUPPE

An *Angelschaluppe* from Helgoland at around the turn of the twentieth century.

Another river boat, but one found more specifically on the River Elbe, was the *buttjolle* – again literally a 'sole yawl', soles being plentiful here in the river. These double-enders had a flat sternpost with an outside hung rudder, a rounded and high stem and very wide planking. Up to the mid-nineteenth century they were about 5 metres long, with those from the lower part of the river a bit bigger and with a square-sail on a single pole mast. Unusually, this had two reefs in it which were at the top of the sail and not the bottom. The fishermen had two masts for their boats, with the second, or extra, one being called a *stormmast* which, needless to say, replaced the larger mast in adverse weather.

After the mid-part of the century, half-decked *jolles* had a forecastle and a wet-well amidships. Like most other German craft, they were completely built of oak with massive frames. The lug rig was introduced in the 1880s, or earlier, with this being a somewhat high sail for the boat size. Transom-sterned *jolles* appeared in the early twentieth century.

The *buttjolles* were mostly used to fish with tangle-nets by one man and a boy, although other forms of fishing were practised. A little boat was towed to enable them to get onto the mudbanks where the nets were set so that they could empty them. Today, only one such *buttjolle* has survived, and this has been converted to a motorboat.

Nine-metre *butt-kutters* developed into this century after the larger cutters, of which we will discover later. These were found to be cheaper to repair by having narrower planks. *Helena*, HK453, is one such cutter that, certainly until recently, was still afloat. These craft have also been termed *zeeg*, from the German *Ziege*, meaning goat, although these seem to have been sloop-rigged.

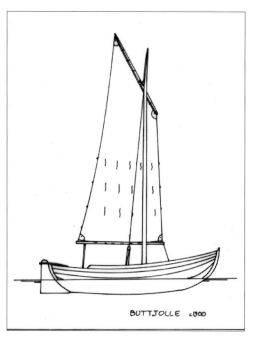

Above right: A *buttjolle* under sail on the River Elbe with the lugsail from a small support boat set as an additional sail.

THE *EWERS*

The *ewer* – pronounced 'ava' - literally means a 'boat driven by one man'. Somewhat similar to the *buttjolles*, but flat-bottomed, they were one of Germany's most interesting and well researched fishing boat types.

In 1750, the small town of Blankenese perched itself upon the northern bank of the River Elbe. Although nowadays it is lost amid the sprawl that is Hamburg, 250 years ago it was the fishing centre of the lower Elbe. The Danes ruled here until the mid-eighteenth century, and the small rowing *ewers* that were moored off the small beach were strongly influenced by these overlords. Soon they adopted a squaresail (*rahsegel* in German) and became known as *pfahlewers*, or literally 'polemast *ewers*'. These were used for river fishing and sometimes to venture out a little into the wide and wild North Sea. Later still, they were used as pilot boats for the growing ports of Hamburg and Cuxhaven. And it wasn't long before the squaresail was replaced by the lugsail, probably around 1830; like the *buttjolle*, this was very high, narrow and square-headed, though.

The *ewers* were double-enders, very pointed and hard-chined with a heavily raking sternpost, but they had a flat bottom and a *kahn* plank – see diagram. Their layout was not unlike a Dutch *botter* – possibly coincidentally, maybe from a direct influence – in that they had a forecastle and a wet-well amidships and used narrow leeboards.

On the other hand, the shape of the hull is said to have evolved from the Danish (Norwegian) punt, which was a very basic three-plank boat. The hull was raised up, with the original three-plank part becoming a wet-well, and the boat's shape developing into what we now regard as the *pfahlewer*, with a carvel-built hull entirely of oak. Indeed, some similarities are obvious between British cobles, and other Saxon boats for that matter. But surely that is how craft evolved in time, through the taking ideas from region to region, adopting them for their particular usage, and passing on other innovations? And with transient fishermen from all over northern Europe drifting after the same fish, the eventual coming together of ideas was an obvious result.

Ewers were supposedly excellent sea boats, although this is now contested by some who doubt this was possible. After stories of their sea-keeping abilities, one of these early *ewers* was replicated and, so it appears, she didn't quite behave as was expected when she went to sea.

By 1860 the town of Finkenwerder (or Finkenwarder in old German) that lay on the south side of the river, nearly opposite Blankenese, had become the main fishing centre. The main fishmarket was at Altonaer, which was Danish until 1860 – the people of Hamburg

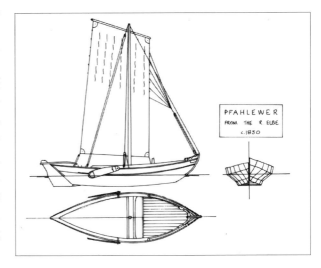

PFAHLEWER FROM THE R ELBE c.1850

describing these inhabitants as being too near to their own town, 'too near' being the literal translation of Altonaer! The Finkenwerder fishermen concentrated on using the new trawls that had only recently appeared. For this they preferred the gaff rig, so they altered their sails and became known as *giekewers* (the *gieg* being the main boom). They then adopted a second mast in the 1870s, both in their desire to sail further afield and for more power for bigger trawls. Then came a bowsprit with its added foresails. The hull at the same time became softer in the chine until it eventually became round-hulled. Yet they still retained the protruding bottom bit and the *kahn* plank. Transoms appeared, giving better buoyancy aft. However, although suited to trawling, the *ewer* was said not to be built strong enough, especially by way of the larger wet-well (*bunn* in German). Several were lost by breaking up in heavy seas because of this.

Ewers weren't just confined to fishing craft, in the same way as the *schaluppen* weren't. *Fruitewers* carried pears and apples from the Alte Lande (old land to the west of Hamburg) to the Hamburg market. *Stonewers* brought bricks from the kilns in the Alte Lande to satiate the huge expansion of the city. *Galeasewers* were large cargo-carrying *ewers* while *stintewers* fished solely for *stint* – similar to smelts and regarded as the poor man's fish.

Ewers fished all manner of ways. Some stowboated like the English smacks, while others drifted for herring in the summer season when the crofters of the inland Westfalia traditionally joined in at Finkenwerder. Later on in the year, eels were the sought after prize.

Numbers of *ewers* in the Finkenwerder fleet show their popularity, and subsequent decline. From forty-two registered in 1850, this grew rapidly to sixty-five in 1860, seventy-two in 1865, eighty-six in 1870 and 110 in 1875. But by 1880, although the fleet had risen to 176, most of these were the new cutters, as we shall discover.

Today no fishing *ewers* have survived, save for the '*Maria*', HF31, which is now in the Deutsche Museum in Munich and was originally built in 1869. A replica was built in 1978 and currently sails in the Baltic.

One final word about their building before we move on. The bottom was built first and caulked, then, after being turned over, it was supported at both ends and weighted in the middle to create the rocker effect. Frames were then fitted into place and cross-pieces fixed to act like floors. Each alternate plank was then bent into shape and nailed on and the structure left to dry out before the fixing of the other planks. Decks, strengthening and gear were then added before the vessel was rigged and launched. A truly remarkably shaped vessel!

THE DEVELOPMENT OF THE KUTTER

In 1865 four deeper draughted English fishing smacks were brought in by the fishing companies, and another twelve subsequently built locally to the same design. Ernst Dreyer, a naval architect, at the same time took the lines from a Grimsby smack and built

Bremerhaven with a collection of fishing craft, including a decked *giekewer* on the extreme right.

Above left: Giekewer 'Maria', LL56, under full sail in little wind. Judging by the lines fore and aft, the boat is moored and posed for the camera. (*Photo: Altonaer Museum*)

Above right: An Altenwerder *giekewer*, ALT243, in 1957, showing its underwater section and the flat-bottom. (*Photo: Gerald Timmermann*)

seven similar vessels. Another seventeen of these were built locally. This meant that there were almost forty English-type smacks fishing out of Finkenwerder within a few years.

All these companies soon went bankrupt and the boats were sold back to England. Whether it was the boats themselves that caused this sudden insolvency, or poor fishing, is unclear, but it does seem that the design was totally unsuitable as they were too deep for the river, too heavy and too expensive to run, which perhaps suggests the former. Forty vessels unable to fish would be a major headache for any fishing company and certainly would put them to the wall, so to speak. However, the impetus to alter the design surfaced again, so that, in 1878, Ehlert-Kuhl from Blankenese produced the prototype *ewer-cutter*. Although this still retained the *kahn* plank, it had a small keel and centreboard, and these became called the *kiel-ewers*. Within a couple of years cutters with bigger keels and no centreboards then became known as *kutter-ewers*, and these tended to have a round-shaped hull and did away with leeboards for the first time. Thus these were a true compromise between the *ewer* and the English smack. Later on, the counter stern was adopted, and within another two years came the final demise of the *ewer* with the true cutter – a more upright, flat-floored, round-hulled boat with a ketch rig, a proper keel and, for the first time also, no *kahn* plank whatsoever. Hence the generations-long traditional shape of the *ewer* finally disappeared. These cutters, with hardly any rise of bilge, and about 9 metres long, became known as the *Finkenwerder kutters*.

Incidentally, before the *ewers* finally disappeared, many of them were lost in the gales of the North Sea as competition from the growing band of steam trawlers forced them to fish in winter. However, so did many cutters for the same reason, and these tended to capsize in the waves and sink while the *ewers* didn't. They merely drifted out of control, to finish up wrecked on the leeshores. This then goes to show that they were seaworthy vessels after all, and perhaps the builders of that replica mentioned just didn't quite get it right. Perhaps!

Three boats on the River Elbe. There are two *giekewers* and one deep-sea cutter between. Sails and nets are drying while behind several other vessels are moored.

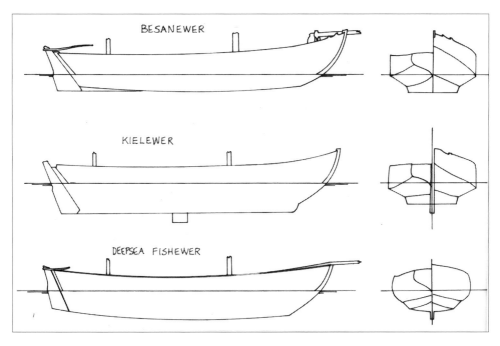

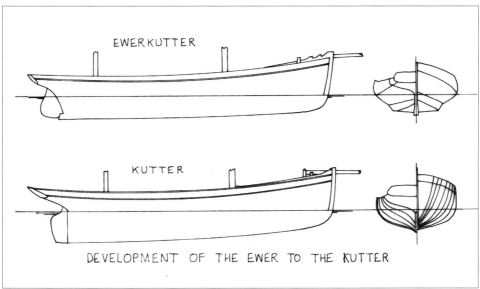

The hulls of the *Finkenwerder kutters* were always painted black with white moustaches, not unlike Scottish craft, although with bigger moustaches. They worked the *hochseefischerei* – high sea fishery – trawling throughout most of the year and drifting for herring in the season. Similar cutters worked out of Cuxhaven. With motorisation they were found to take units well and many survived the ravishes of war.

Today several remain, all in different states of repair. On a visit to the area I found the 1925, Tonning-built, '*Lily Hilda Johann*', which was only decommissioned from fishing in 1984 and which has now been converted for pleasure. Then, in the back garden of a friend with whom I was staying, I came across an older HF-registered

The round-bilged *ewer-kutter Germania*, HF203, being scrubbed. She was built at Finkenwerder in 1886.

boat that he is hoping to soon completely restore. Some other surviving examples are the recently-rebuilt '*Landrath Kuster*', HF231, and the museum-owned '*Catarina*', HF287, while the '*Providentia*', HF42, is currently under restoration and the '*Ora et Labora*' was sitting ashore in Lanzarote awaiting attention.

KRABBENKUTTERS

Krabben are shrimps from the healthy fishery previously mentioned in the shallow waters all along this coast. In the sixteenth century, herring was king here. Taxes were paid on fish landed on the island of Sylt, at Germany's extremity along its northern border with Denmark, and where the epicentre of this herring fishery was to be found. In 1500 there were some 220 rowing boats from the island fishing within sight of Helgoland, that tax-free haven some 30 miles offshore. Twenty years later, this had increased to 340 boats. This, remember, was at the time that the German Hanse merchants controlled the fishery of Sweden's south-west coast. No doubt they brought influence to bear upon these fishers.

But, by 1607, the buss fishery had established itself. These busses had a crew of forty-five men and there were some fourteen ships engaged. However, these have been referred to as large *ewers*, a tenuous description indeed. Four years later, only four had survived, with small, open boats – *fischer-jolles* – in control. A century on and these small *jolles* had acquired a square-sail, and by 1771 there were thirty such craft working out of Sylt.

In the next century, the *kutter* was adopted, much as the English smack was introduced on the River Elbe, and this seems to have coincided with the re-emergence of the shrimp fishery. Maybe one supplanted the other. The waters around the islands between the mouth of the Elbe and Sylt are shallow, and perfect breeding grounds for the shrimp. These craft, influenced by the Danish cutters from the north, became known as the *krabbenkutters* or 'shrimp cutters'. They were single-masted gaff cutters of up to 10 metres in length, and were built at harbours such as Tonning and Busum. After motorisation, the first one being engined in 1906, these cutters were easily adapted and, although their shape altered somewhat in fullness, they retained a certain similarity as before. The local yard at Tonning built some 190 motorised cutters between 1920 and 1959, when the last one was built. This was the '*Pornstrom*', TON4, which today sits ashore as a silent reminder of the hundreds of these craft that were built. Several sailing cutters remain afloat today.

Today shrimpers work out of Tonning, Busum and Friederickskoog, but of course these are modern boats that bear little resemblance to the cutters, though, as we've seen, similar cutters worked all along the Danish coast.

A 1970s postcard of *krabbenkutters* at Neuharlingerseil.

Below: Similar, though newer, *krabbenkutters* at Fedderwarderseil in 1994.

THE HERRING FISHERY

As previously mentioned, the herring fishery quickly established itself as the main North Sea fishery, and was controlled by the large buss-fishery up until the eighteenth century. Here it was known as the *Große Heringfischerei* and was as important economically for the Germans as it was for the Dutch and British. Emden, on the River Ems, was wholly dependent on it, and here, in 1820, there were twenty-seven busses. However, the first French-type *lougre* arrived in 1867 and the design was instantly a success and copied. Vegesack, on the River Weser, became the main base for these *loggers* after about 1895. Over the next 20 years there were some fifty *loggers* working from Vegesack, and these were motorised after about 1906. These boats were similar to their Dutch counterparts and a more detailed description will be found in the next chapter. One fine example does

Steam trawlers alongside at Cuxhaven, *c.* 1910.

remain, and the aptly named '*Vegesack*', BV2, built in 1895 by Bremer Vulcan at their Bremen yard and restored in 1979, still sails out of the port under the ownership of the society 'Maritime Tradition Vegesack', chartering during the summer.

Steam overwhelmed the German sailing fleets, causing the same decline here as elsewhere around the North Sea. Bremerhaven became the prime steam trawler port, with nearly 200 steamers in its fleet. Cuxhaven and Ems had substantial, if not as large, fleets until two wars and overfishing caused the massive decline in fishing seen throughout Europe, albeit perhaps worse here than elsewhere, possibly a result both of the defeat in two world wars and the physical distance of Germany from the Atlantic fishing grounds. What was left, however, of the history of their fishing fleets fuses itself with that of the rest of the North European fleets from this point on.

SOURCES

There are numerous books on the traditional vessels of Germany, all written in German and mostly concentrating on the deep sea and herring fisheries. Thus we have:*Logger-Jantjes* by Wilfried Brandes, *Die heringsfanger von der Mittelweser* by Manfred Scheller, *Ein Schiff wird kommen* (the history of the sailing logger *Vegesack* BV2) by Tham Korner and Gerald Sammet and *Seegekehit & Seegesalzen* by Gerhard Kohn. For an overall picture, Gerhard Timmermann's *Handbuch der Seefischerei Nordeuropas* studies the craft throughout North Europe. *Ewer Maria* by Jobst Broelmann and Timm Weski is a superb study of the 'ewers'. There are, of course, numerous other books on the subject and there are also several museums, including those at Hamburg and Bremerhaven, which include fishery-related artefacts. Cuxhaven has a fishing museum. *Piekfall*, the magazine of the 'Freunde des Gaffelriggs' – literally 'friends of the gaff rig' – has been publishing a wealth of material since 1973.

CHAPTER 9

The Netherlands

*The River Ems to Antwerp,
Including the Zuiderzee*

Holland has, for many centuries, looked to the sea to expand, discover and fish. Much of its expansion, as well as colonisation in the Orient, involved reclaiming land from the sea, for much of the country lies below sea level. Thus this 'low country', as we discovered in north-west Germany in the previous chapter, is reliant upon dykes and waterways that have made its people masters of such construction. Not surprisingly, the North Sea has played a major part in the development of the Netherlands' maritime links, so essential

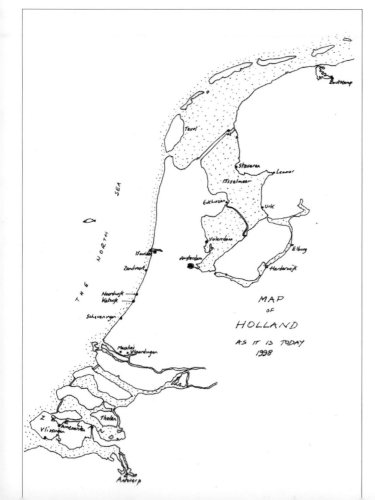

in earlier times. Its fisheries were contributing much to the Dutch economy long before the other countries around the North Sea were realising this natural resource lying just offshore. This was the herring fishery, which Dutch fishermen and curers developed by 'inventing' the drift-net and mastering the curing of herring in salt. However, almost as important, was the inshore fishery on the varied expanses of the inland waterway system, from the sheltered waters of the Schelde in the south to the Waddenzee (Frisian or 'shallow' sea) and the great Zuiderzee in the north.

In the previous chapter, mention was made of the bluff-shaped *Norderneyer schaluppe*, similar vessels to which worked in the Dutch waters of the Waddenzee, many being built around the coast of the Zuiderzee that was open to the sea until the twentieth century. Other craft include the *visserspink* and *schuyt*. The former were the most common inshore craft between 1400 and 1800. They worked the coast from Egmond on Zee to Zaandvoort, Noordwijk, Katwijk and down as far as Scheveningen, at what became the port of the capital of Den Haag. Although originally double-ended, very pointed craft, they gained a much fuller shape, becoming in time transom-sterned to increase deck space. They were sprit rigged with a boomless sail and worked both the inshore herring and long-lines. At the same time, the *schuyt* were working their drift-nets fairly close inshore. One particular type – the *Heijester schuyt*, a two-masted square-sailed vessel with forecastle and wet-well – was working off Katwijk in the eighteenth century. Similarly, a *bezaan schuyt* fished off Pernis, near Rotterdam, in 1752 with a Katwijk crew. Fitted with leeboards, these vessels are said to be the forerunners of the *bomschuit*, one of the most remarkable types found in Dutch waters – if not in European waters. The *bezaan schuyt* is also said to be the originator of the mizzen *bezaan* sail – a mizzen sail suspended from the sprit and boomed aft.

THE BOMSCHUIT

The outstanding common feature of the beach-based *bomschuit* was their 2:1 length/breadth ratio. This gave them a squat, squarish appearance, and on deck a *bomschuit* really was rectangular with rounded-off corners. Their traditional home was the beaches of Scheveningen and Katwijk, where fishing, especially in the latter, has been intertwined with everyday life for centuries. Records of Katwijk mention that in the fourteenth century the fish market was moved to the sea village, where the old church had been built in 1480. Fishing, though, remained exclusively a poor man's trade until a thriving smuggling trade developed in the seventeenth century. Fishing only became profitable after the lifting in 1857 of the Monopoly Act, which had previously given the right of fishing solely to square-rigged keel boats, thereby excluding all flat-bottomed beach boats and foreign-owned vessels from catching herring. During this prohibition, however,

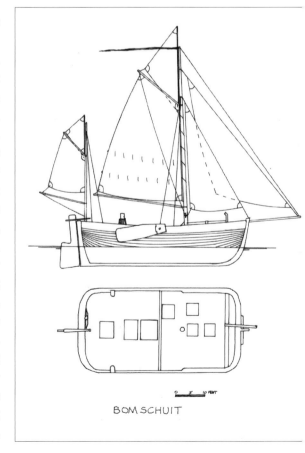

BOMSCHUIT

A *bomschuit* under full sail heading for the beach, where horses will drag her ashore. Four knots was about the best they could make.

A *bomschuit* is jacked up before manpower and horses will swing it through 180 degrees to face the sea once again.

the villagers largely ignored this unjust law by catching, gutting and salting their herring to preserve them for selling outside the village. They had to be subtle in their catching ways.

The clumsy shape of the *bomschuit* – which literally means 'bluff boats' – did account for their sluggish speed, although they certainly weren't as unstable as they might appear at first. They managed about three or four knots under full sail in a good wind, and were mostly gaff-rigged with two masts, both having topsails, and foresails. Looks are often misleading, though, as the *bomschuit* often sailed north to join in with the spring herring fishing. Photographs show various examples at anchor with the Scottish herring luggers and the like at Lerwick in Shetland.

Their construction was said to be an art, though why any more than other boats is unclear. Surely all boatbuilding is art. However, they were built in the main at one of four yards at Katwijk. The most prolific of the builders was one W. Taat, who had two covered yards in the town – one for building sharp *boms* and the other for full *boms*, for different owners had their own preferences, which accounts for the sometimes acute eccentricity of the shape. A *bom* was simply a smaller version of the *bomschuit*, a single-masted vessel used for shrimp fishing, though the actual difference is very blurred, so that both terms appear regularly intermixed. The smaller *boms* had a maximum length of 28 feet and a breadth of 10 feet 6 inches but otherwise their construction was exactly the same as their big sisters.

CONSTRUCTION OF A BOMSCHUIT

The oak bottom was first laid out much in the same way as a keel is laid for a keelboat. This consisted of three oak planks, 12 cm thick, being laid beside each other and fixed to 20 cm by 20 cm transverse oak members, or *kespers*, using 1.5 cm pine nails. Then the oak stem and sternpost were mounted and affixed. Planking the hull – the most labour-intensive part of the process – could then begin. Moulds were used at either end, but only the experience and eye of the builder could create the middle part of the hull. Elm planks were used below the waterline, and these were bent into shape in the burning hold, or *brandgat*, where burning wood pulp had copious amounts of water added to prevent sparking and the consequential igniting of the planks. The hull was then built up, with the planking supported from the ground by uprights. The 5 cm planking was riveted to the previous plank using 6 cm galvanised iron square nails. Above the waterline, Bavarian or Slavonian oak of the best quality was preferred. Once the planking was complete, the frames and floors could be added, then the deck. Building time was eight to ten weeks for a 50-ton hull, with eighteen shipwrights working. Cost at the turn of the century was 11,000 Dutch guilders.

The beaching apparatus was unique in Europe as far as is discernible. Various techniques such as turntables or greased timber poles have been used; the *bomschuit* used a corkscrew attached to the boat at its top. This jacked one end of the vessel up into the air. Using horses and manpower attached by ropes to the other end, the vessel was dragged around through 180 degrees so that it was again facing the waves. The hammering the boats received meant that their working lifespan was relatively short.

A storm on the coast in 1894 destroyed some twenty-five *bomschuit* from the Scheveningen fleet and damaged another 150. This gave an incentive to the authorities to build a harbour there, which led to the demise of these extraordinary craft. The *logger-bom* – a flat-bottomed boat with a *logger*'s bow and stern and a length/breadth ratio of 3:1 – was already beginning to gain preference, so that once the harbour was completed, the transition was abrupt. One of the last *bomschuit* to be built was the *De Jonge Leendert*, launched by W. Taat in 1891. At about 14 metres long, the boat could carry some 340 cran (280 barrels) of herring. A crew of ten manned the vessel – skipper, mate, cook, four seamen, a boy, lookout and line-shooter – all of whom lived forward, with the net store and fish hold aft. On this vessel three men were expected to share a berth, though on some craft there were only two to a bunk. In 1905, *De Jonge Leendert* had a keel added, although this meant that the boat could not be beached. Like so many others, she then worked out of IJmuiden, where the canal system could be entered from the sea. Because of this link between canal and sea, IJmuiden grew up into one of the Netherlands' main fish ports.

By 1910 the days of the *bomschuit* on the beaches of Katwijk and Scheveningen were past. However, those days when the beaches were a hive of activity have, fortunately, been captured on canvas. Katwijk was home to many painters and among them, and best of all, was Hendrik Willem Mesdag, whose great illusionary masterpiece *Panorama Mesdag*, in Den Haag, where the painter created a 360-degree

The *bom* KW88 as a static exhibit outside the lighthouse in Katwijk in 1997.

panoramic seascape showing the beach at Scheveningen as it was in 1881, still exists and is open to the public.

Only one *bomschuit* has survived. The KW88 is actually a shrimp *bom*, sometimes also called a *garneschuit*, and she sits outside the lighthouse in Katwijk as a reminder of the hundreds of these unique craft that once existed just across the road on the beach. The KW88 was built in 1911, which was quite late, and she ended her working days at the Zuiderzee Museum at Enkhuizen. Brought then to Scheveningen, the Katwijkers reacted angrily on seeing her SCH registration number. Within a year the two communities agreed that Katwijk was her true last resting place, and she returned there several years ago.

THE ZEELAND COAST

The southern part of the Netherlands, Zeeland, is a maze of river estuaries and islands. Around these waters, and down as far as Antwerp in Belgium, where it was called the *Flushing mussel boat*, the distinctive *hoogaars* is said to be a centuries-old craft, although its exact roots are unknown. Up to the nineteenth century, *hoogaarzen* (plural) were used for all manner of work – mostly freight, ferrying and fishing – but by the late 1800s this was exclusively the working craft of the shrimp and mussel fishers.

The *hoogaars* is flat-bottomed and can sit on the mud at low tide, yet the planks have an extreme lift at either end – more at the fore end, giving the boat its most extreme feature of a high bow with a heavily raked stempost. This characteristic helps produce an exaggerated sweeping line of its rubbing strake. Various alternatives characterised vessels from different locations along the coast. The northerly boats from Duiveland had virtually completely flat bottoms, while those from Arnemuiden on Zeeland itself were renowned for being deeper draughted – being deepest at a point directly below the mast and curving up to merely a few inches at the waterline fore and aft. This, it has been said, made them tend to broach in heavy seas, whereas the fuller-sterned Tholense *hoogaars* was much more seaworthy because of the greater buoyancy aft. The *kinderdijkse hoogaarzen* – inshore boats with hardly any draught – were ultimately sold on and many went to work at Lemmer on the Zuiderzee.

The *hoogaars* was originally a sprit-rigged vessel, having no boom to optimise the working space. However, the high gaff sail on a short, curved gaff – for which Dutch boats are well known – was adopted by many hoogaarzen fishermen in the nineteenth century. Paradoxically, the Arnemuiden boats generally retained the sprit and many of these adopted the *bezaan* sail. Some *hoogaarzen* had a wet-well, while most had a shrimp boiler. All had leeboards – not surprisingly, as nearly all the Dutch craft had leeboards. The boats were built of oak on oak and consequently had a long life, although the harsh working conditions of continually scraping along coarse banks perhaps shortened this. The area around Zeeland literally teemed with these vessels up to the end of the nineteenth century, yet nowadays only a few have survived as working vessels. In common with many other Dutch traditional working craft, however, the *hoogaars* design has been adapted for the pleasure yacht business. This has resulted in many of these traditional types being built in modern materials, a development which presumably has received mixed blessings.

Another traditional Zeeland type was the *hengst*, which also worked the mussels and shrimps. These flat-bottomed, *bezaan*-rigged craft, with a transom stern, were indeed similar to the *hoogaarzen*. One particular *hengst* was reputedly used in 1711 to ferry Johan Willem Frisco, from which he drowned. A variation of the *hengst* is the *lemsterhengst* – a vessel with the bow of a *hengst* and stern of a *lemsteraak*

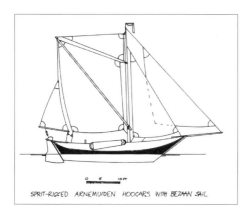

SPRIT-RIGGED ARNEMUIDEN HOOGARS WITH BEZAAN SAIL

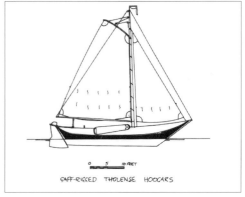

GAFF-RIGGED THOLENSE HOOGARS

An Arnemuiden *hoogaars*. Note the patching on the spritsail, and the 'roller' furling.

A gaff-rigged *hoogaars* from Vlissingen. The gaff is long and the topsail unusual but effective in the narrow, confined waters of the south in which they worked.

(see below) – which was used by the mussel fishermen to import mussel spat from the Waddenzee into Zeeland.

The *schouw* was a centuries-old design used mostly for inland seine-netting. Larger, heavier-built, transom-sterned *schouw* also worked out to sea off the Zeeland coast. Again, these were originally sprit-rigged, becoming gaff-rigged in the latter part of the eighteenth century. The smaller

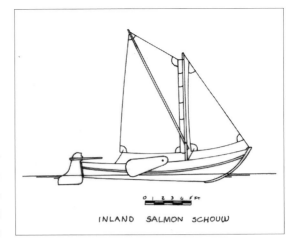

INLAND SALMON SCHOUW

river or *salmon schouw* retained the sprit rig. The type is particularly recognised by its rounded profile fore and aft, which supposedly makes them easy and cheap to build and probably accounts for their abundance throughout Zeeland, the Zuiderzee and Friesland.

THE DEEP SEA

The ancient deep-sea craft of Holland appear to have been the *zeeschuyt*, two-masted, square-sailed craft that are the earliest documented form of herring drift-net boats. Similar craft – the *buizen* or Dutch busses – followed within a century, and these larger craft sailed out to Shetland in early May and chased the migrating herring shoals southwards until Christmas. Being at sea for months on end, they had to be stout and capable vessels which were in turn served upon by smaller *jacht* that sailed back and forth, bringing supplies and taking barrels of cured herring home. The *dogger* was a form of the *buis*, a ketch-rigged vessel that trawled in the seventeenth century. It is said that many of these *doggers* worked the North Sea, especially in one shallow area where huge amounts of fish were found, and that this area became known as the 'Dogger Bank'. Later Dutch trawlers working the same area were also referred to as *doggers*.

Closer to home, the *hoeker* long-lined for cod and haddock offshore. Larger versions – about 66 feet long – were substantially built and sailed up and down the English Channel and as far away as southern Ireland, where they are said to have influenced the Irish in some of their fishing boat designs. Similar sloops also worked out of Belgian waters. When the Monopoly Act was repealed in 1857 there was, not surprisingly, a sudden opening up of the herring fishery to all. This coincided with the superseding of hemp nets by the lighter cotton ones. Overnight the herring fishery became a bonanza for small craft that were previously excluded from taking part.

A few years later, in 1865, a French *lougre* was introduced into the Scheveningen fleet and was immediately deemed successful, so that more were quickly brought in.

A Vlaardingen *logger* being posed for a photograph. These reached the peak of their development in the 1890s, about the time this photo was taken.

The 1910-built *logger 'Lotos'*, pictured here in 2008. She fished until 1929 and then, in 1939, being the last sailing *logger* in the Holland, she was sold. She moved to Norway, where she worked as a coasting vessel until returning to Holland in 1993 under private ownership.

The 1916-built *'Gallant'*. Launched as the *'Jannetje Margaretha'* from the shipyard of the Figee Brothers in Vlaardingen, she fished for herring until moving to Denmark. She returned to Holland for restoration in 1987 and is now used for sail training.

The last *hoeker* disappeared in 1886, the last *buis* already having gone. Over the next 20 years some 400 *loggers*, as they became known, were built in Holland, mainly around Rotterdam. The herring fishing flourished once again among a keen home market and a healthy German export, with the port of Vlaardingen becoming the

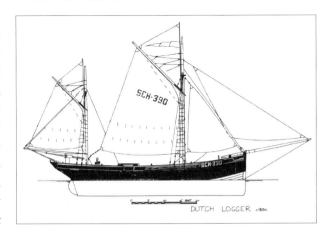

DUTCH LOGGER c1830

major herring port. With the building of the railways, this hit a peak towards the end of the nineteenth century and into the next. Steam capstans appeared towards the end of the nineteenth century, so that by 1903 nearly every *logger* had the added advantage of being able to set longer trains of nets. At the same time steam drifters entered service, but these seem not to have had much effect upon the building of new *loggers*, and they continued to be built up to the First World War when Holland retained its neutrality, hence ensuring a continuance of the thriving herring fishery. The last working sailing *logger – Dirk*, KW44 – was sold to Germany in 1931 for conversion into a motor sailing coaster. However, several examples have survived and sail within today's Dutch charter fleet such as the 1910-built *Lotos* and the 1916-built *Gallant*.

THE ZUIDERZEE

Although the inland sea of the Zuiderzee is a relatively small area, it has a long coastline, some of which adjoins the North Sea, so some of the vessels already mentioned worked within its waters. However, given its size, it has itself a rich variety of fishing vessels in its own right. A series of books by Dutchman Peter Dorleijn contains comprehensive descriptions of both these vessels and the folk who operated them. The only intention here is to attempt to clarify and characterise the many types.

The Zuiderzee is literally the 'South Sea', as against the North Sea that we know. On 28 May 1932, this waterway was finally closed to the sea by the completion of the long dyke across its northern entrance. Salt water then eventually became fresh water. Until that momentous time, fishing had been the primary source of a livelihood for the thousands living close to its shores. Large shoals of herring and anchovy were prevalent in these shallow waters as they came to spawn year after year. Suddenly they were prevented from achieving this. Virtually overnight, fishing patterns changed within what became known as the IJselmeer, and to a lesser extent outside as well. Before the dyke was ever dreamed of, the herring, and particularly the anchovy, were the most profitable of the catches in the spring for the fisherfolk. During the summer and autumn they fished mainly for flounder, eels, smelt, shrimps, mussels and

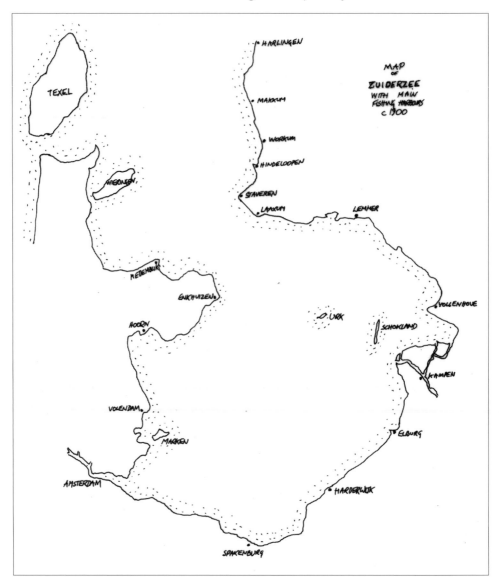

periwinkles. In winter the fishing was largely non-existent as the waters were often frozen over.

At around the turn of the century, there were some 1,000 large wooden craft fishing the Zuiderzee. At first glance these all appear similar: that is bluff, with full-bodied shapes and a certain clumsiness about them. They were designed specifically to sail on the shallow water, which was only on average 2–3 metres deep, and up and down the canals, lowering and raising their rig many times a day. Yet, on closer inspection, there are so many subtle characteristics making up different configurations that it really does seem like these indigenous craft were unique among European fishing craft. To characterise them is initially difficult, but most do fit into one of two categories – either round-hulled or hard-chined in section, though there are the few that fit somewhere in between.

BOEIERS AND BOTTERS

The *boeier* is one of the oldest of Dutch craft and is said to be the generic Zuiderzee boat, with graceful lines and lapstrake construction that reflects its seventeenth and eighteenth-century origins. It has a full, rounded bottom with curving stempost and fairly straight stern. It often had a wet-well, was some 9 metres long and was in common usage up to the mid-nineteenth century.

Similar to the *boeier* is the *botter*, perhaps the best known of all Dutch craft. This developed over the nineteenth century, mostly around the west and south parts of the Zuiderzee. Some say it evolved from the *zwak*, although these tended to be merely bigger versions of the same. The *botter* is basically a combination of the hard-chined, sharp *punter* hull with the *boeier*'s rounded lines. It is an open boat with a high bow and foredeck, and is regarded as an exceptional sea boat, especially in the short, sharp seas of the Zuiderzee. Most of them working in that water were about 15 metres in length, although 18-metre versions worked out into the Waddenzee and even into the exposed waters of the North Sea. They were gaff-rigged with the characteristic curved gaff boom. Having the mast set well aft, they had a large jib and foresail on a long bowsprit. Like the *boeiers*, many set a *bezaan* sail from a halyard set aft on a high bumpkin or boom called a *bezaantutter*. These were handy craft that worked well to windward. By the end of the nineteenth century the *botters* were in use throughout every corner of the Zuiderzee, and they remain the most prolific craft to have sailed these waters. Another feature of the *botter* is the narrowness of the leeboards, probably due to their good performance. *Botters* survived through the motorisation era, with their hulls being particularly receptive to engines, so that by 1920 there were still 146 registered as *motor botters*. Today many pace the Dutch coast and further afield, some being original while others are yacht-like replicas. They still remain as one of the best all-round craft to have surfaced from the Dutch waterways.

The other early vessel appears to have been the aforementioned *punter* – or *zeepunter* – which is regarded as the archetypal Saxon craft of the Zuiderzee. It was a hard-chined, flat-bottomed wooden craft, with sharp ends, and was generally completely open. Like most of the Zuiderzee craft, it too was gaff-rigged although the *togenaer*, a barge-like half-decker which was similar, was sprit-rigged. It is probably true to say, however, that most of the early *punters* were sprit-

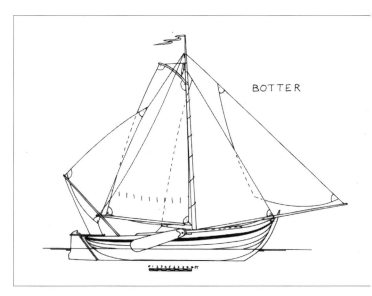

BOTTER

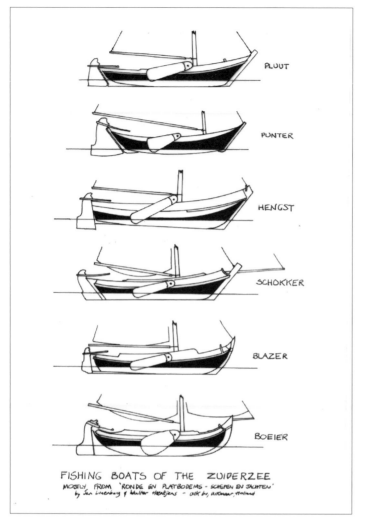

FISHING BOATS OF THE ZUIDERZEE
MOSTLY FROM 'RONDE EN PLATBODEMS - SCHEPEN EN JACHTEN'
by Jan Linenburg & Walter Hentjens - Alk by, Alkmaar, Holland

Below: Botters
at a shipyard in
Sparkenberg in 1900.

1. The old church converted to the fisheries museum at Hel in Poland with boats exhibited outside.

2. The big Pomeranka boat at Hel, DEB5, with straight ends and a straight rudder configuration.

3. A Nordlands boat is seen in northern Norway with either agricultural goods or timber, showing that they were not exclusively used for fishing. Here a topsail adds to the powerful rig and again there is a *lofting*. The small boat in the foreground is a *tororing*.

4. A Nordlands boat sailing in a good breeze between Brest and Douarnenez, Brittany, in 2008.

5. Clinker-built, gaff-rigged *sjöbåtar* – sea boats – off the Swedish coast in a painting from 1885 by Jacob Hagg.

6. Modern-day craft at Galtaback, western Sweden, in 1996. Today's craft reflect the ancestry of the past.

7. A redundant fishing boat on the coast of Latvia. (*Photo: The Latvian Institute*)

8. The *kurenai*, the boat of the Curonian Lagoon.

9. Redundant motorised beach boats lying on the shore of the Bay of Puck, Poland.

10. Boatbuilder Jacek Bielecki working on his new canoe at the Vistula Lagoon Museum, Poland. This was a replica of one from the 1940s. The boat was to have two spritsails, was flat-bottomed and was made up from two pine (*stepka*) with planks three planks either side. Other than the bottom planks, the rest was oak.

11. The inside of the Nautineum Museum at Dänholm on the German Baltic coast with a wonderful array of canoes and *heuer* craft.

12. A *Kleinkutter* – or small inshore fishing boat – on display at the Nautineum Museum.

13. Two boats sailing off Brittany from Brest to Douarnenez in 2008 – the bisquine 'La Cancalaise' on the left and the 'Tina Husted', a later breed of Danish fishing boat, on the right.

14. The Icelandin sailing trawler 'Sigurfari', the last boat built at Burton Stather, Hull, in 1885 as the 'Bacchante'. She was sold to Iceland in 1897 for £325 and her name changed about 1900 to 'Sigurfari'. In 1919 she moved to the Faroes, where she fished until 1970, thereafter moving to the Akranes Museum Centre, Iceland, where she remains. (*Photo: Jennifer Snell*)

15. The *'Prasident Freiherr Von Maltzahn'*, a 22.8-metre *Hochseekutter* (deep-sea cutter) at Hamburg in 1996. She was built in 1927 by J. J. Sietas at Cranz-Neuenfelde and fished until 1969.

16. The herring logger *'Vegesack'*, BV2, off the coast of Denmark under full sail in 1994. (*Photo: Tham Korner*)

17. Dutch artist Hendrik Willem Mesdag's view of the fleet leaving. Note the vessel on the right has a small mizzen mast.

18. Part of the masterpiece *Panorama Mesdag*, showing the beach at Scheveningen in The Netherlands as it was in 1881.

19. Drive into Oosduinkerke, Belgium, from the motorway and this boat stands at the entrance to the village as a silent memorial to those who lost their lives fishing.

20. The top of a tin of biscuits showing a fanciful Victorian scene with a *Bourcet-malet* rigged beach boat being admired by holiday-makers at Berck, northern France.

21. The *bisquine* '*La Cancalaise*' making an impressive sight sailing with the flotilla off Brittany from Brest to Douarnenez in 2008.

22. A motorised lobster boat from Tazones, northern Spain.

23. The *galeaos 'O' Abandonado'* in 2008. Built in 1916 as a fishing boat, by 1926 she was coasting along the Portuguese coast and later internationally. The boat sank in 1968 and was raised years later for restoration. She now charters out of Noirmoutier, France.

24. Colourful boats drawn up at Camara de Lobos, Madeira, in the 1970s.

25. La Linea beach, southern Spain, in 1999, with several rowing traditional boats used for the seine-net (*jábega*). The men to the left are using one of the old wooden capstans to haul a boat up the beach.

26. 'Seascape at Saintes-Maries' by Vincent Van Gogh. The boats are *nacelles*. (*Photo: Pushkin Museum, Moscow*)

27. The Cinque Terre ports in Italy are full of traditional boats, crammed in between the high buildings of the harbour which almost seem to hang from the mountainside.

28. The fishing boats were all motorised by the end of the Second World War, although, judging by the way the sternpost has been cut away, this was a later addition to this boat.

29. The *Gozo boat 'Maryanne'* in Gozo in 1995.

30. Today's fishing craft of Croatia are similar to those of Southern Italy and Greece, though they are still equally pleasing to the eye. This is Dubrovnik. (*Photo: Jan Pentreath*)

31. A *trechandiri* from the west side of Corfu. Although most of these craft have some form of (usually ugly) superstructure, the hulls between the Ionian and Aegean Seas have subtle differences in shape.

32. Two boats selling their catch at Kamena Vourla, Greece. Though this is technically illegal these days, according to the nonsense from the European Union legislators, some fishermen thankfully continue the age-old practice of selling direct to the customers.

rigged in the nineteenth century and before, as were many *boeiers*. The *punters* were common along the south-eastern part of the Zuiderzee, from Lemmer to Elburg. Likewise, the *pluut* is another type common along this portion of the coast, specifically between Harderwijk and Elburg. Similar to the *botters*, these developed in the second half of the nineteenth century from early sprit-rigged craft, and lasted throughout the first half of the twentieth century. The last one was working up until 1946. Like the *punters*, they were flat-bottomed with a 45-degree, raking, yet straight, stem and sternpost. They had a rounded hull form, however, rather than a hard-chined one. Today several examples of both types remain in service.

SCHOKKERS AND BONS

The fisherfolk of the islands of Urk and Schokkerland (now part of the *polder* or 'new land') and, to a certain degree, Enkhuizen, developed their own craft in the eighteenth and nineteenth centuries and these became known as *schokkers* for obvious reasons. They were flat-bottomed, hard-chined craft and were said to be one of the finest sea boats of the Dutch tradition. Many fished out into the North Sea waters and around the Frisian Islands. Most were half-deckers at about 16 metres long, while there were some larger versions which were fully-decked at about 25 metres. All had thick, straight stems and sternposts which both raked to about 45 degrees, and most had a wet-well to keep the fish alive. By the twentieth century they had become rounded in shape and a little fuller in the bow, and had long, straight bottoms. They appear to have originally developed from the *pluut*, although some refer to the *hoogaarzen* as small *schokkers*. More likely, influence for the development of the *hoogaarzen* came from the *punters*, as there are many similarities. Furthermore, there are also similarities between the *schokkers* and *botters*, although the latter are a bit finer. Given, however, that all these craft work in the same proximity, similarities in characteristics are to be expected.

A *schokker* from the small town of Het Bildt on the Friesian coast under full sail.

The *schokkers* were renowned for their strong rig of a boomless main, *bezaan* sail and jib. This made them suitable for working around the Frisian Islands, where communities relied mostly upon fishing for their living – apart from the summer, when there was a certain amount of tourist trade for sea bathing. The constantly shifting sandy waters also gave rise to a fair amount of salvage work. Several *schokkers* remain today. The *bons* – otherwise known as *Vollenhove schokkers* – were quite similar in shape to the *schokkers*, albeit a little bit smaller at 10 metres and a bit sharper at the stern, giving them a more pronounced 'S' shape at their point. These were common in places such as Elburg, where the last remaining example was restored a few years ago.

AAKS AND BOLS

The *aaks* or barges were much fuller than the finer-shaped *punters* and *schokkers*. *Aaks* are basically ships without a stempost and in general simply have a front plank, which then gives them their rounded appearance. Various differences between *aaks* from different parts of the coast came about through the nature of their usage. The *lemsteraak* from the east was far more rounded in the bottom than the flat-bottomed *wieringeraak*, a heavy, large vessel developed in the second half of the nineteenth century around Makkum and Workum. *Mosselaaks* were, as their name implies, barges used for the collecting of mussels. *Enkhuizenaaks* were a little wider than the *lemsteraaks* which, as we've seen, influenced the Zeelanders in the design of their *lemsterhengst*, and the *boeieraaks* were nineteenth-century Zeeland barges used alongside the *hoogaarzen* for mussel and oyster cultivation. *Boeieraaks* were more bluff-bowed than any of the other *aaks*. At Dokkum, on the Friesland coast, a replica *wierumeraak* has recently been built as a memorial to the eighty-three male members of the population of the small villages of Wierum and Paesens Moddergat, north of Dokkum, drowned during a particular storm on this coast in March 1883. This is a two-masted gaff-rigged boat with a wet-well and these vessels are said to have sailed out into the North Sea as far as the Dogger Bank to fish.

Lemsteraaks (as they are called in the Lemmer dialect rather than the more widely used version of *lemmeraaks*) were the favoured boats of Friesland and were said to be the fastest of all the Zuiderzee craft. Lemmer, however, was only home to a group of part-time fishermen who fished largely inside the Zuiderzee, in contrast, say, to those of Urk, who were regarded as having families involved in fishing stretching back generations. Today this is apparent in that there are only two or three boats working from Lemmer but some twenty-five inshore craft at Urk, alongside the same number of North Sea boats.

These fishing *aaks* were about 11 to 12 metres in length as larger boats were considered too heavy for fishing. *Eel-aaks* were 18 to 21 metres and much fuller and heavier with flat bottoms. These had big wet-wells to transport the eels, which were bought off the fishermen around Greestmaar and Heng and stored ashore before being

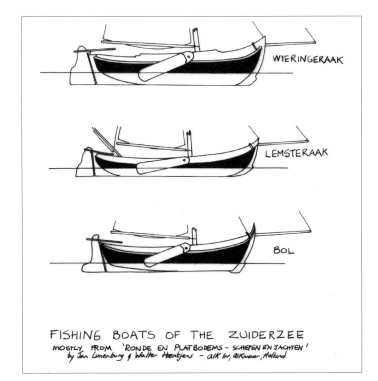

FISHING BOATS OF THE ZUIDERZEE
MOSTLY FROM 'RONDE EN PLATBODEMS - SCHEPEN EN JACHTEN'
by Jan Limenburg & Walter Hoentjens - alk bv, alkmaar, Holland

Below: Lemsteraaks unload herring at Lemmer

loaded aboard the vessels and transported away to market. London was the chief market and one Lemmer boat was said to have been constantly moored up alongside the fishmarket at Billingsgate selling eels, another arriving to replace one after having exhausted its supply.

The *lemsteraaks* always have the *bezaan* sail boom fixed to the main boom – unlike the botters and many other Zuiderzee boats, which have it set further aft. This ensures the sail works efficiently when the boat is on a beam reach. Furthermore, the fishermen never used a foresail set on the bowsprit when sailing on the wind, only setting up the smaller jib. Otherwise, they believed, this gave far too much sail upfront and unbalanced the vessel.

Lemmer had a fleet of 120 boats in 1895, which seems strange given that the fishermen were part-timers. Either they were rich enough to be able to build a boat or the owners came from outside Lemmer. This, however, doesn't explain how the name of the boat came about. By 1910 the number of boats there was in decline. Today only one original wooden *lemsteraak* remains, the *Zevija*, LE39, built in 1898 at a cost of 1,800 guilder and which was based at Harderwijk a few years ago.

Steel *aaks* appeared around the beginning of the twentieth century, with *Presto*, LE50, being built at Lemmer in 1901 by A. de Boer, a descendant of a boatbuilding family particularly renowned for their fine craft. She survives today under the ownership of a 'board', complete with her original Lister two-cylinder 7.5 hp engine, which was fitted in 1952. She has a traditional layout of two bunks in the fore cabin, stove and nets aft of this cabin and a wet-well and large working space. At 12 metres,

Above left: The *lemsteraaks* were the favoured vessel of Friesland and one of the fastest Dutch craft.

Above right: This *lemsteraak*, LE72, powers through the water off the harbour at Lemmer.

The *lemsteraak* LE21 flying its distinctive *bezaan* sail.

The *lemsteraak Presto*, LE50, the first steel aak in Lemmer.

she was larger and finer than the earlier wooden versions. Today *lemsteraaks* are still built in steel – though only as pleasure boats – and upwards of 100 craft join in with the local racing.

The *bol* is another of the old Dutch craft in use before the *hoogaars*. It is a round-hulled craft, wider and squarer than many other types, and has barge-like qualities which make it similar to the *wieringaak*. There was a *wieringenbol*, a smaller, around 25-foot, vessel built at Makkum, Workum and Hindeloppen for flatfish and shellfish fishing. It appears that neither of these two types were built on Wieringen, which was an island in the Zuiderzee at one time, in the north-west, although it became attached to the mainland because of reclamation. The *bol* generally originated from the town of Vollenhove and one, RD23, was seen in Harderwijk a few years ago, though there are more afloat.

BLAZERS AND KUBBOOTS

Blazers were the successors to the *schokkers*, and were heavily-built craft that worked both inside the Zuiderzee and outside of it. Some even sailed as far as Zeeland. Like the *schokkers*, they were gaff-rigged but were more rounded like the *botters*, whose influence was largely responsible for their arrival in the late nineteenth century. They were most common in the north of the Zuiderzee and among the Frisian Islands. As well as being more heavily built, they were lower and fuller in the bow, lay better in

A small *kubboot* used for catching eels, here at Harderwijk.

the water, and the majority were never completely decked over. Furthermore they, too, had wet-wells and set a gaff main and jib only, much in the same way as the *botter* did. Today the *blazer De Drie Gebroeders*, TX11, is a replica of one built in 1894 at Makkum. One of the largest ever built, at about 14.5 metres by 5 metres, she was completely decked, contrary to the norm. She was one of the first boats fitted with a motor, receiving a 16 hp unit. Today this vessel can be seen sailing around the IJselmeer or at the Zuiderzee Museum at Enkhuizen.

Kubboots were small open boats built before 1925, specifically for fishing eels on the west and south coasts. The name derives from *kubben*, which were traps consisting of wicker baskets filled with bait and hung up on poles in the water just above the seabed. These traps were emptied daily by the fishermen in their *kubboots* – often towed by larger craft to the vicinity, or rowed/punted. Some *kubboots* had a small mast with a triangular mainsail that was laced to the mast to enable it to be constantly lowered by letting go a single halyard. Some sailing types set a small foresail. Most worked the eastern part of the Zuiderzee between Marken and Vollenhove.

STAVERSE JOLS

Staverse jols were originally small, bluff-bowed open boats about 6 metres long that were used for the herring and anchovy fisheries with a drift-net or long-line, specifically from Staveren, on the east side of the Zuiderzee, and the surrounding area. These *jols* were the only Dutch craft with a full keel and transom stern and consequently had no need for leeboards. They, without doubt, evolved from a definite Scandinavian

influence and this is further evidenced by their use of the word '*jol*'. Having a deeper hull enabled them to work the deeper waters to the north and around the Frisian Islands. They were developed from the need to have no projections overboard while fishing the drift-net. Larger *Staverse jols* of about 7 metres developed in the early part of the twentieth century, some having a fo'c'sle, while the smaller types used a canvas cover to act as the same when working away from home for any extended period. Today several original examples remain, one being at the Zuiderzee Museum.

Several years ago at Laaksum, just to the south of Staveren, the bow portion of the *jol* HL6 was sitting

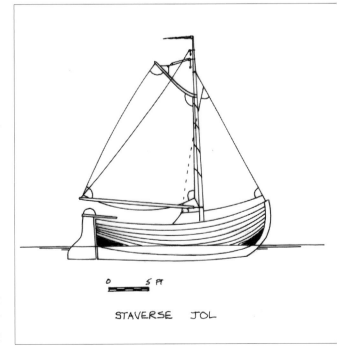

STAVERSE JOL

Above left: The distinctive bluff-bowed *Staverse jol*, here pictured in a garden.

Above right: Unloading anchovies from a *Staverse jol* at Medemblik.

as a reminder of this way of life that has now altogether disappeared. There used to be some twenty *jols* based in the tiny harbour before the *dijk* closed the Zuiderzee. Now the place is empty, save for a few small boats, and this really is typical of the IJselmeer – the whole way of life of a rural community has vanished within a couple of generations. By the time the section of the boat rots away – maybe it has already – the memories of this way of life will too cease to exist.

There were other, less common craft that operated in the waters: the *boatsje* was a small, barge-like river boat; the *Mariakerker jol*, a sprit-rigged small inshore boat of about 6 metres from the Belgian Schelde region; the *vleet*, a small, tender-like, high bowed, round-hulled boat common throughout the Netherlands; the *grundel*, a small inland water boat originating from the mid-nineteenth century; the *zomp*, a wooden canal barge that carried fish into the hinterland of both the Netherlands and Germany; and the *sloep*, an inshore cutter-rigged boat up to about 50 feet that was generally the result of a mix of Belgian, British, German and French influence. The *grundel* was found mostly in the peripheral regions of Zeeland and Groningen as early as 1890, but it was never successful in comparison to Dutch shallow-draught vessels. Dutch craft are without doubt plentiful, and their mastery of the sea throughout the globe is still apparent. Their skill at the fishing nets preceded that of Britain, and it was to the Dutch that the British looked in the development of the British fisheries, especially that of herring. The variety of working craft reflected this determination to conquer the seas, and the result has been a diverse collection of wooden boats from within a relatively small length of coastline compared to many other European countries. Of course, as elsewhere, motorisation affected the design and numbers of traditional craft, many of which disappeared from use. Iron and steel construction, the latter for which the Dutch are highly regarded today, superseded the use of timber and eventually the shape of fishing craft changed drastically so that they no longer had much resemblance to older craft. However, with a healthy number of restored and replica craft still sailing, probably more so than any other European country, there are many who still strive to keep at least some sense of the old traditions alive.

SOURCES

There are a whole host of books on Dutch fishing vessels, though very little written in English. Of the former, the books by Peter Dorleign are the most comprehensive, with details of the vessels, the fisheries and the fishing communities. The National Fisheries Museum at Vlaardingen has an extensive resource while the outdoor museum at Enkhuizen is internationally known for its exhibits. Other, smaller museums exist and, more importantly, the Netherlands has an extensive fleet of traditional craft which can be seen sailing throughout the summer months.

Traditional vessels are very much alive in Holland, often racing in the narrow waterways.

CHAPTER 10

Belgium

Antwerp to Dunkirk

Sandwiched in between the bulk of France and the richness of Holland, Belgium often appears forgotten by those of us skimping through the maritime influences of the north European coast. How different it is in reality. Although the Belgian North Sea coast is relatively short – only 45 miles – Belgium is one of the most densely populated countries in the Europe, so it's hardly surprising that they have had to use this natural resource to its full extent.

Nowadays the coastline is a heaving mass of touristy towns reaching from the French border at De Panne right up to Knokke-Heist in the north – ostensibly the posh end of the coast! However, driving along this coast, without seeming too unkind, none of it appears particularly up-market, but simply an endless sprawl of hotels and bright lights resembling many Mediterranean coasts, where many of these resorts would perhaps be better suited rather than facing what is often a grey, blank North Sea.

But the coast is sandy hereabout, which presumably attracts the punters, magnificent beaches that shelf gently so that huge expanses of sand are exposed at

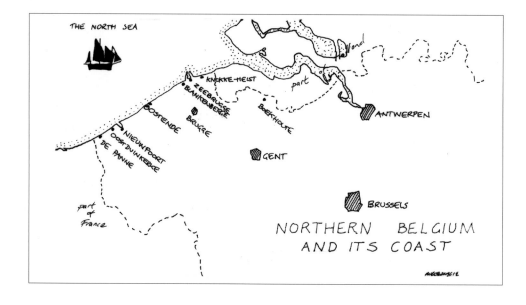

THE NORTH SEA

Holland

KNOKKE-HEIST
ZEEBRUGGE
BLANKENBERGE
OOSTENDE
BRUGGE
NIEUWPOORT
OOSTDUINKERKE
DE PANNE

BOEKHOUTE

ANTWERPEN

GENT

part of France

BRUSSELS

NORTHERN BELGIUM
AND ITS COAST

low water. These are backed by dunes onto a low, flat Flemish hinterland. Perfect, yes, for holiday-makers and fishermen alike. One hundred years ago, these resorts were the workplace of the Belgian fishermen. Today, like elsewhere, fishing is concentrated in a few harbours and consists mostly of deep sea vessels with uncanny capabilities of completely denuding the seas of all fish, although it must be said that the country only has a fleet of about 100 vessels. Licences, like in Britain, have been sold to their Dutch neighbours.

The majority of fishermen worked off the beaches before the twentieth century. The principal bases, with their port registration letters, were: De Panne (P); Koksijde (C – due to the old spelling of Coxyde); Oostduinkerke (OD); Nieuwpoort (N); Oostende (O); Blankenberge (B); Zeebrugge (Z); and Heist (H). On the River Schelde there were several centres of the fishing. Antwerp (A) and Bouchoute (BOU) – now called Boekhoute – were the most important, while there were other ports of minor importance along the river such as Baasrode, near Bouchoute; Doel (D); Kieldrecht (K); Lillo (L); Mariekerke (M); Ruppelmonde (R); and Zandvliet (ZV).

The earliest records tell of the rowboats used between 50 and 150 years after Christ, but details are vague until the tenth century, when the squaresailed craft – like the Brugges ship – were prevalent. These formidable boats are generally regarded as being the first *hoekers* and herring boats that worked inshore. By the thirteenth century the *dogger* or *dog boat* was working further out into the North Sea. In the following century it was reported that at Scarborough there were 110 boats from Flanders, thirty-one of which came from Nieuwpoort, thirteen from Lombardsijde, twelve from Oostende Mariakerke and eleven from Blankenberge.

Nieuwpoort is the only natural harbour along this exposed coast. The harbour at Oostende, however, does date back to the fifteenth century. And the situation remained thus for years, with all the other fleets working from beaches, in the same way that the Dutch and French – and British for that matter – did. In more recent times, Blankenberge's harbour was built about 1875, and that of Zeebrugge around the turn of the century. De Panne was destined for a harbour as well until war broke out in 1914, after which all these plans were abandoned when the Armistice had been sealed.

By the sixteenth century Dutch-type *pinks* were working alongside the *hoekers*. Herring busses appeared at the same time, these being very bluff three-masters. Again like the Dutch, these were served upon by fast *jagers*, bringing the catch to market. Similarly, three-masted *fluit* sailed to Greenland to catch whales. In the eighteenth century *bunhoekers* – hoekers with wet-wells – first appeared.

But the true Belgian – Flemish, really – fishing boat was the *schuit* (*scute* in 1500, *schuyt* in 1700 and *schuit* in 1900). In the fifteenth century these boats worked out of Blankenberge, Oostende and Heist. A painting of 1425, now in Nieuwport, shows 'scuten' (plural). The Blankenberge *schuit* of this era was 11.5–12.5 metres long and had four or five crew. Inside the double-ended hull was a forecastle with accommodation. The vessel would have been rigged with two lugsails, the foremost of which was set on the mast in the eyes of the boat that raked forward. In true Dutch fashion, these luggers retained their leeboards into recent times. Similarly, the Heist boats (Heist being the main fishing station in the north until Zeebrugge was built) were double-ended, very bluff and flat-bottomed, yet they nearly always had a cutter rig with topsail. Oostende, on the other hand, had keeled *schuiten* to suit the harbour. These, too, set a gaff sail yet they had a transom, and appear very much as the boats from the south of the coast. Other Oostende *schuiten*, probably those working off the beaches of Mariakerke and Middelkerke, were flat-bottomed, double-ended and lug-rigged. And, just to confirm the mix of north/south flavours, we find that Nieuwpoort and De Panne luggers all had transoms, while, as we've already seen, those from the north were all pointed. It always seems to be a matter of refining the best of both worlds between Dutch and French attributes.

The *schuiten* originated from the Viking *skuta* after their coming here in the twelfth century. Although these were fast double-enders, the theory is that the boats soon became flat-bottomed to work directly off the exposed flat beaches while retaining their double-ends. So the *schuit* evolved. They were used for all manner of fishing, from drift-netting for herring and setting bottom nets to trawling by anchoring and

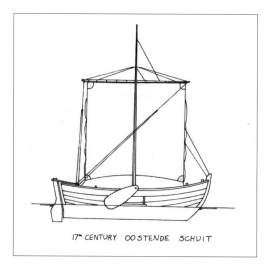

17ᵗʰ CENTURY OOSTENDE SCHUIT

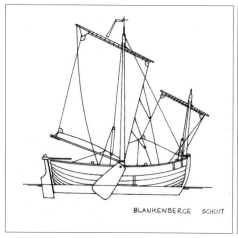

BLANKENBERGE SCHUIT

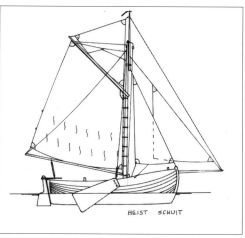

HEIST SCHUIT

pulling themselves forward using the capstan. They were derisively nicknamed 'toads' because of their squat appearance. The last working boat was built in 1915 and this, along with the another fifty-odd vessels working there before the First World War, was destroyed by the Kaiser, along with another sixty that worked off Heist's beach. The '*Vryheid*', B17, is believed to have been the very last vessel to fish.

One 12-metre replica *schuit* was launched in 1999. The brainchild of, among others, Daniel Bosser, a former fish salesman, this vessel, the '*St Pieter*', named after the patron saint of fishermen, took seven years to reach fruition. Drawings for this had been obtained by taking the lines off a model of such a vessel that had been commissioned by the Church and still existed in the Blankenberge townhouse. The backbone of the replica vessel is oak, the keel being 460 mm wide and laid flat before the 85 mm by 85 mm doubled frames were fitted. Planking is of 450 mm by 25 mm elm with some extreme bending at either end. For this a fire was built alongside a steel framework so that, by throwing water on the hot plank continuously, it is subjected to a constant force around the metal cross-bars and hence into shape. Each plank took a whole day to bend and fix. Transverse strength comes from three massive thwarts across the vessel while in the forepeak are four low bunks, copying those the fishermen lay in while lying to their drift-nets.

The Blankenberge *schuiten* had four bottom planks laid side by side in carvel fashion, followed by five clinker planks. The Heist versions were clinker-built throughout, this being another of the differences in the characteristics of each type. One remarkable factor in all the *schuiten* was the fact that, for a 12-metre vessel, they were 5.5 metres in beam and only 60 cm in draught. Two stout oak leeboards create

The *schuit* '*St Pieter*' under sail leaving
Blankenberg in 1999.

the necessary stability. The rig, as mentioned, consists of two dipping lugsails set on vertically-stepped masts. In the fifteenth century the forward mast was usually raked forward in its place in the eyes of the boat, though over the centuries this mast became more upright. Modern requirements demanded an engine which was fitted with some awkward planning. The propeller emerges very close to the waterline, which makes it inefficient, though doesn't impinge on the general appearance of the vessel. In light winds it is surprising how well the '*St Pieter*' sailed along at three or four knots.

De Panne luggers – called *Panneschuiten* – are indeed more than tainted with a French flavour. These three-masted boats that worked off the beach do most definitely conform to characteristics from further west along the coast, as shall become evident in the next chapter. With two lugs and a tri-sail mizzen, complete with topsails and jib, these clumsy looking craft were worked as recently as the 1930s. Two-masted, gaff-rigged, *Panneschuiten*, though retaining the mizzen tri-sail, first appeared in the eighteenth century from, it has been suggested, English influence. Working off the beach for shrimps, the gaff wasn't really suitable, and few fishermen were persuaded of its advantages. The gaff rig was preferred by those working out of Nieuwpoort

Two Blankenberge *schuiten* moored off the beach. These were lug-rigged vessels with leeboards and flat bottoms, with a strong resemblance to the beach boats of Katwijk, Holland. (*Photo: Belgian National Fishing Museum*)

A Blankenberge *schuit* on the beach, showing its bluff stem with a fair amount of curvature and an extremely rounded stern. Note the size of the boat in comparison to the fisherman in the foreground. (*Photo: Belgian National Fishing Museum*)

harbour. All these boats were built locally at De Panne. At the turn of the twentieth century there were two builders still working – Messrs Mols and Denye. The '*Scharbiellie*', P1, which survives outside a church in De Panne, is generally referred to as a *Pannesloep* and these were single-masted gaffers though the open hull was similar in shape to the bigger boats. These small boats did work directly off the beach, fishing mostly for shrimps in competition with the men on horseback that dragged trawls through shallow water.

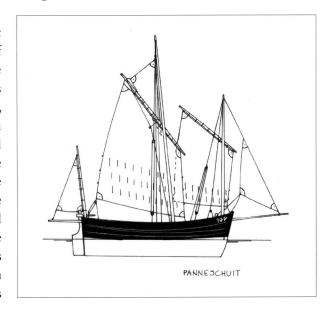

PANNESCHUIT

The shrimp fishery was the most profitable of Belgian inshore fisheries, the small crustaceans liking the sandy coastline which, as mentioned, shelves slowly into deep water. Second to this was the catching of soles. Indeed, the fishmarket at Oostende very recently was awash with both of these, and little else. Shrimps were traditionally caught using a trawl towed behind a horse in some parts, as was commonplace in parts of Britain, most notably Flookburgh on Morecambe Bay. On the southern part of the Belgian coast it was practised by fishermen/farmers in their off season as far as the land was concerned, and mules were the norm for the dragging of the trawl through the shallow water. The method continues today, more as an attraction to inquisitive tourists than a way of life, but, nevertheless, several beefy fishermen, upon the sturdy horses they favour these days, work the beach at Oostduinkerke during the months of May, June, September and October one hour each side of low-water. During the height of the summer season there are usually too many jellyfish to make the trawl worthwhile, though they persevere for the tourists' benefit.

At Oostende another particular type of fishing vessel was used for shrimping – the *schover*. This type developed over the second half of the nineteenth century out of the flat-bottomed beach boats. The *schover* was an 8-metre boat, clinker-built with a transom stern and single-masted lug sail and foresail set on a long bowsprit. These shallow-draughted craft were unique to the Belgian shores, although similarities can be made between them and some of the French beach craft, and perhaps even some of the beach boats from Suffolk, England. Similar shrimpers worked out of Nieuwpoort.

Sloops were introduced into the Belgian fleet in the 1800s, probably from British influence. These were used mostly for trawling or long-lining in the North Sea from about 1865. Oostende had its own fleet of gaff-rigged *longbomer* (longboomers) and even Antwerp had substantial numbers of home-based sloops. Many of these sailed as far as Shetland, the Faroe Islands or even Iceland with a dozen or so crew. Many of these became dandy-rigged *kotters*, as they were called, with a boomed main, topsail,

Here the full shape of the *Panneschuiten* is appreciated. The three-masted rig was common among Dutch, Belgian, French and British craft up to the late nineteenth century.

Above left: The *Pannesloep* 'Scharbiellie', P1, which was a single-masted gaffer used to shrimp-trawl in competition with the fishermen working on horseback.

Above right: The 'Scharbiellie' today, surviving as a permanent exhibit outside a church in De Panne.

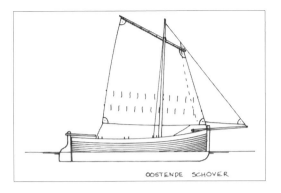

A typical horse-borne shrimp trawl fisherman with his net and panniers for the shrimps. I wonder what the fellow behind is in such a hurry about?!

Schovers at Oostende, boats used specifically for catching shrimps. Although originally lug-rigged and clinker-built, they became carvel-built with a gaff in the 1930s.

The harbour at Oostende with *schovers*, all motorised by this time.

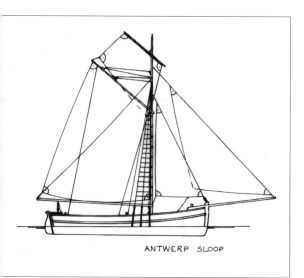

ANTWERP SLOOP

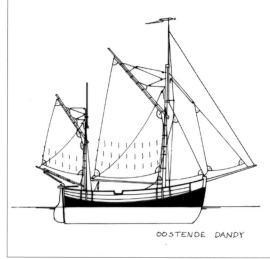

OOSTENDE DANDY

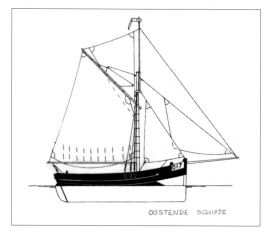

OOSTENDE SCHIPJE

staysail, jib and mizzen, and sometimes a mizzen topsail. Some sloops were used for herring drifting and stowboating for sprats.

British smacks were introduced into the fleets after the First World War. Many Flemish fishermen stayed in England during this war, and learned to understand the smack and appreciate its qualities as a fishing vessel. Consequently, many were bought up after the war and sailed home to fish from Ostende and Nieuwpoort. These dandy-rigged vessels were called *Oostendse Dandies*. The replica 'Nele', built in 2005, runs charters from Oostend.

In 1895 a prototype of a new craft was introduced from Norway into the Belgian fleets – the *schipje* was a 13-metre vessel that first worked out of Ostende for shrimping in deeper water, and was somewhat similar to the bawleys of the River Thames. These cutter-rigged boats, often painted with a large moustache, are perhaps more reminiscent of the German cutters of the Elbe in outward appearance, although, as shrimpers, totally unlike the Elbe craft in their work.

One ketch-rigged shrimp boat, built in 1939, is the '*Jacqueline Denise*', B72, which first fished out of Oostende as O72 before being decommissioned in the early 1950s and becoming a pleasure craft. By 1991 she was languishing in Blankenberge. She was restored and handed over to the 'Scute Association' that had built the '*St Pieter*' and now sails out of Blankenberge.

The River Schelde, which borders the north coast, was home to a small mussel and shrimp fishery. Here *hoogars* of a typical Flemish or Dutch design worked the shallow, tidal waters. Villages such as Paal, Kruispolderhaven and Emmadorp were once home to substantial shrimp fisheries that disappeared with the reclamation of land projects. The Dutch had wanted to dyke across the entrance to the river in the same way as they had on the rivers to the north, but Belgian insistence seems to have prevented this. However, large tracts of land have been altered and restrictions put upon fishermen. Today all the remaining *hoogars* are Dutch, and these traditionally raced each July to the Antwerp market to land first, thereby attaining the best price, to satisfy the Belgian desire for mussels. It seems they are the favourite dish in the north of the country. Bouchoute was a mussel centre, and the mussel spat was collected from Nieuwpoort's

Oostende *schipjes* in the harbour. These were only introduced into the Oostende fleet in 1895 and have been said to have been modelled on the bawleys of the River Thames estuary.

Oostende smacks – or *loggers* – sailing out of the harbour in about 1910. Note the steam tug towing these vessels out into deeper water.

An Oostende *logger*. These were modelled on French and Dutch boats and introduced in the second half of the nineteenth century.

pier and brought here to be seeded in the river. Other traditional Belgian fishing craft were the river-working yawls of Mariekerke and Hamme (River Scheldt) and Lier on the River Nete.

Loggers in the French style worked from the deep harbours in much the same way as they were introduced into the Dutch fleets. As we've already seen, these were two-masted luggers that adopted a ketch rig and worked all along the French, Belgian, Dutch, German and Danish coasts, out into the North Sea and beyond. Steam first arrived in 1884 in the form of the steam trawler '*Prima*' at Oostende. The following year saw the second steam trawler, '*Elisabeth*', and the days of the sailing boats were finally numbered here too, as they were throughout northern Europe. Boats adapted to motor power in the second decade of the new century, as elsewhere, and the steamers and loggers gained preference offshore. The flat-bottomed beach boats survived perhaps longer here then than in many other parts, probably because of the shallow water, which required such craft. The *schover* survived well into the twentieth century, but before long the ubiquitous motor trawler superseded all other craft everywhere. Technology, unfortunately, cannot be suppressed at the expense of conservation, and the resulting chaos in the European fishery was perhaps only a by-product of years of mis-management. The Belgian fishery, however, has probably fared much better within this European framework than have many other fisheries, especially the British. With such a short sea coast and a massive appetite for seafood, Belgium's fishery reflects a determination found upon a fishery of substance and one that has retained its own uniqueness even though it has been heavily influenced by its larger neighbours.

The 36-metre '*Amandine*', O129, was built in 1961 and was the last boat to work the Icelandic waters. She was decommissioned in 1995 and later became a static exhibit in a concrete berth at Oostende. Below decks there is an exhibition on this fishery which is worth visiting. Also at Oostende, visitors can experience the work of traditional shrimping aboard the '*Crangon*', which runs trips out in the summer.

SOURCES

The National Fishing Museum (National Visserijmuseum) at Oostduinkerke has been recently refurbished and tells the story of the coastal fishery. Booklets are also available there on various aspects of the fishing. In summer, shrimp fishermen demonstrate fishing on horseback on the beaches when the tide is right. Furthermore, the Maritime Museum at Antwerp has some information and models of fishing vessels. The '*Amandine*' in Oostend has a superb collection of photographs aboard.

CHAPTER 11

North and West Coasts of France

Dunkirk to the Gulf of Gascoign

France indeed has a coastline of contrasts: from the sandy beaches of Dunkirk, the muddy estuaries of Picardy, the cliffs of Normandy, the rocky inlets of Brittany and the huge sweep of sand and pine trees south of the Gironde estuary to the Mediterranean coast, this coastline must surely rank as the most diverse in Europe. To match this diversity, there have been some sixty different types of fishing boat catalogued as fishing from these rivers, lagoons, beaches and ports in the times before motorisation, a diversity matched only by Britain. There are, of course, more but these are generally variations of some of the main types. With a coastal length of 2,130 miles, the majority in the north and west, lying facing the English Channel or the Atlantic seaboard, it is this part of the coast that this chapter concerns itself with. Here there are some fifty craft types identifiable and for convenience the coast has been split into four areas. Here, more so perhaps than in most of the other European countries, many communities have forged links with their maritime past by building replicas of the vessels their ancestors once relied upon. Some original vessels also exist, although many fishing

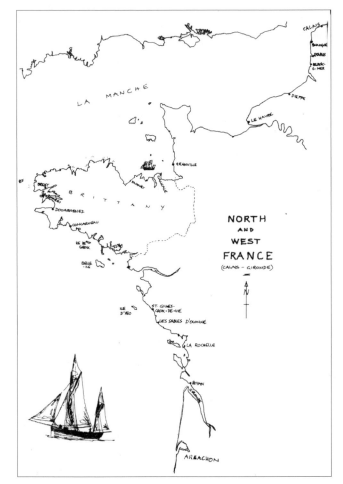

NORTH
AND
WEST
FRANCE
(CALAIS – GIRONDE)

boats were destroyed both during the German occupation and, as elsewhere within Europe, subsequently through national and European governmental policies.

The influences that brought about the designs of the vessels that once fished off the coast of France, in the English Channel (*La Manche*) and around the northern part of the Bay of Biscay, are the same ones that led to the evolution of many of the boats of the south and west coasts of England. What is immediately noticeable is the flow of influence from both north and south. From the north come the techniques of the Vikings and Saxons with their clinker beach boats, which are squat and tubby like the Dutch and Belgian beach craft. From the other direction come influences from the Veneti and the Spanish. They meet somewhere in the middle! Inter-mixed with these influences are the peculiar needs which are wholly dependent on individual preferences and the local conditions.

NORD PAS DE CALAIS

Three-masted luggers characterised all the fishing boats between Flanders and Normandy in the eighteenth and nineteenth centuries, such as those seen at La Panne in the previous chapter. Both Dunkirk and Gravelines had their own versions of luggers which fished during the nineteenth century. In Dunkirk the boats became known as *Grenadiers* from the particular grey shrimp, the *garnel*, that they fished. Both versions of the lugger were similar in hull shape, this being full in body, steep in the dead rise and flat-bottomed without much of a keel, perfect for being beached in the tidal estuary. They both set lugs on the main and foremast with a main topsail, though in Gravelines they preferred a small lug mizzen (*tapecul*) whereas the Dunkirkers set a small triangular mizzen. Both needed a good breeze to sail well off the wind, and they were never regarded as being close-winded craft.

The hull shape suited the shallow entrance to Gravelines and the short seas of the North Sea and Channel. They were built of local oak and were up to some 10 metres in length and seldom worked more than 20 miles from home except when drift-netting for herring, at which time they ventured as far as the Scottish coast, or mackerel fishing in the Thames estuary. At other times they trawled for shrimps, long-lined and set hand-lines. In the last decade of the nineteenth century, larger versions were built to fish the Icelandic cod, these being up to 15 metres in length.

In Dunkirk the luggers were also beached. These also followed the herring and mackerel out of the prawn and shrimp season. These were cooked immediately on arrival back in port and sold as far as Lille and into Belgium. Slightly larger vessels, at up to 12 metres, trawled for flat-fish and set lines for mackerel. By the beginning of the century some adopted a cutter rig, though their use declined. A few survived into the 1930s and early 1940s, at which time they had been motorised, which in turn enabled them to boil the shrimps at sea on their way home.

However, in general, it was the *Dundée du Nord-Pas-de-Calais* that superseded the luggers towards the end of the nineteenth century. The last lugger at Gravelines had

Gravelines sloop '*Saint-Jehan*', a replica of the shrimper that last sailed in 1938. The 1942-built Gravelines shrimper '*Christ-Roi*' has recently been restored and runs charter trips. '*Christ-Roi*' is ketch-rigged with a transom stern.

Above left: A square-sail fishing boat from Calais, very bluff and clumsy, dating from the second half of the eighteenth century. From the look of it, it's an obvious ancestor of the beach boats found between Holland and Normandy, including the south coast of England. The housing for lowering the mast presumes some form of drift-netting while what appear to be sweeps over the starboard side are stowed well out of the way of net activity over the port side.

Above right: A Calais trawler.

been built in 1898, though the first of these *dundées* appeared in 1875. These were much superior vessels, up to 17 metres long and 20 tons, modelled on the Dutch *loggers* and British smacks. They were generally dandy-rigged, which is where their name *dundée* came from – the French pronunciation of 'dandy' and their probable knowledge of the port of Dundee. However, they soon moved the mizzen mast forward to become ketch-rigged. Setting high-peaked lugs and topsails with jib and foresail, they were as powerful as any of the other similar craft working from the North Sea and Channel ports.

In Boulogne they were soon building larger versions, up to 60 tons, that sailed to Iceland to trawl for cod. It is said that they built a dozen each year so that by 1896 there were 105 such vessels in Boulogne. Other ports along the coast such as Le Crotoy, Calais and Oostende were inspired by these craft to build their own, though some preferred to retain the lug rig upon the more efficient hulls.

In the same way as, for example, Lowestoft in England developed both large sailing trawlers and drifters, so did the fishermen of Boulogne after about 1860. Their herring drifters – called *dundées harenguiers* – were ketch-rigged, up to 26 metres in length, upon which much of the herring was cured after catching. By 1900 the boats were even larger, over 28 metres, using capstans to haul their long drift-nets. However, they were soon overtaken by the steam drifters, though some were motorised after the second decade of the twentieth century. Similar *dundées harenguiers* worked out of Fécamp and Saint-Valery-en-Caux, both in Normandy.

But it is the various beach boats for which this coast is better known. Although similar in shape to the beach boats of Brighton (Brighton hog boats) and Hastings across the Channel, there are subtle differences between those of Wissant and Boulogne in the north and the estuaries upon which are the towns of Etaples, Berck, Saint-Valery-sur-Somme and Le Crotoy. The latter, also having a fleet of the larger *dundées*, is tidal and today is the home to several abandoned fishing craft. Today's fleet is moored across the estuary at Le Hourdel, where the *dundées* must once have sheltered.

Wissant is still today home to several *flobarts* (sometimes also *flobards*), the word coming from the Old Saxon 'Vlot bar', meaning 'floatable' or 'buoyant', and relates to the seaworthy nature of these open boats, which are also known more generically as 'Boulogne beach boats'. Though originally clinker craft in Viking style, developed from influences after their settlement of the coast in the eighth century (and suggestions that led to a comparison with the boats of Veneti), later versions were carvel-built in locally grown elm. They had a 2:1 length to beam ratio and were heavily constructed. A daggerboard gave them a grip on the water, for they only had a foot or so of draught for ease of beaching. Rigged with two lugsails, the foremast was stepped in the eyes of the boat while the offset mizzen rested against the transom – the *Bourcet-malet* rig – giving the fishermen a good working area in between. Oars were widely used, especially when launching from the beach. Landing back in was more awkward, dangerous in heavy seas, and upwards of fifteen men were sometimes needed.

Their length ranged between 6 and 8 metres, though smaller versions (4.5–6 metres) appeared at Boulogne and some of up to 10 metres at Equihen which were decked

after the expansion in the availability of salt led to an increased demand for salted fish. At the turn of the twentieth century there were over 150 such craft working from the tidal estuary. The Wissant boats tended to be open, though some fitted a semi-permanent deck when following the herring. After 1925 they were motorised so that today they are still working, although of course some are now fibreglass.

At Etaples the luggers were larger – up to 13 metres in the keel. The estuary suffers from silting and only fishing boats take to the mud on the ebb. Before 1849 there was hardly any fishing, but the arrival of the Boulogne to Amiens railway created a development of wooden quays upon which fish could be landed and whisked away to market. Early clinker-built beach boats were replaced by carvel boats with a pretty sheer

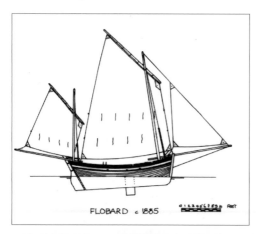

FLOBARD c 1885

line. These had counter sterns and were built by the local shipyards of Caloin, Adams or Brasseur. However, unlike the luggers from the north, they dropped the third mast, preferring a large main lug at over 100 square metres in size, along with mizzen lug, mizzen topsail and jib. Weathervanes with colourful cut-out motifs were affixed to the top of each mast. When fishing, they set a beam trawl over the starboard side which was hauled in with a capstan.

A flobart at Wissant showing its full form.

Being decked over, there were four compartments below: the stern with the steering gear and some stowage and sleeping platforms; the main part, with a coal fire and net and gear stores; the fish hold with ice storage; the forepeak.

Again, boats worked directly off the beach at Berck for inshore fishing. They were flat-bottomed craft again, full in form, with a large transom and, at up to 9 metres in length, they were among the largest beach boats in France. Again, they were rigged with the *Bourcet-malet* rig of two lugsails and were reminiscent of the squat beach boats of Holland and Belgium.

Like many fishing communities, the Berck men had their own disasters at sea. In 1861 five boats were wrecked and thirty-two fishermen drowned. However, by the end of the nineteenth century there were eighty-six boats at Berck, manned by 500 fishermen.

Similar craft at Saint-Valery-sur-Somme were called *sauterellier*, from *sauterelle* – literally 'grasshopper', but referring to the particular pink (mantis) shrimp they fished. Similar in hull shape and rig (except for the triangular mizzen) to the Berck boats, though renowned for their wine glass transom, they were about 8 metres in length and carvel-built in oak or pine. Because of the stony nature of the beach, some used elm in the keel because of its durability when being constantly dragged over the stones. Many had daggerboards.

At Le Crotoy similar vessels fished mostly for herring and were called *etadier*, from the particular gill-net they used – *l'etade*. Again, they had to take the ground and were flat in the floors, full bodied and were rigged with the *Bourcet-malet* rig.

At Berck the boats tended to be larger than those to the north.

Above: Clinker-built *sauterellier* on the beach at Saint-Valery-sur-Somme. These really do sit upright.

Left: A *sauterellier* photographed near Saint-Valery-sur-Mer in about 1992. The daggerboard is raised and the mizzen is displaced to port.

NORMANDY

A boat suited to the local conditions on the beaches of Etretat and Yport, both of which were so-called whaling ports before 1860 (which means local owners sent their ships, crewed locally, to catch whales), was the *caique*, a strongly built boat capable of being constantly beached and put out in surf. Built of elm planking on a slightly rockered oak keel, these boats were clinker-built until the adoption of cut frames brought in carvel planking. They were shallow draughted and rigged *Bourcet-malet*. Generally three sizes were built: small *canot* at under 6 metres, *demi-caiques* at 6–8 metres and the *caiques* of up to 10 metres and 10 tons, some of which were originally three-masted vessels with three lugs and a lug topsail on the middle mast.

One survivor is the '*Vierge de Lourdes*' (F1089), a boat built by Jouen-Fiquet of Fecamp in 1949. This was the last of four boats he built, all similar yet not identical. The others were '*Vive Jesus*' (1947), '*Dieu Protégez*' (1948) and '*Notre-Dame de Bonsecours*' (also 1949). They all belonged to the same family and the '*Vierge de Lourdes*', engined with a 40 hp Lesoeuf unit, was regarded as the best to work, with the most pronounced sheer and a little bit sharper at the bow. She generally line-fished off the Isle of Wight from April to October, while fishing mackerel and herring in the autumn.

The luggers of Le Havre resemble the mackerel drivers of Cornwall while the trawling smacks of Trouville resemble the Essex smacks. Further along the coast from Trouville, towards Honfleur, is the ancient community of Villerville, a village lying atop the hill on the edge of the estuary of the great River Seine. Here, flat-bottomed boats called *plattes* were once pulled up away from the water's edge by wooden

Selling fresh fish on the beach at Etaples, with two-masted *caiques*.

39   YPORT. - *Pêcheurs et Barques de Pêche. LL.*

Above: Again the *caiques* are two-masted, this time at Yport, and there appear to be quite a number of the boats pulled up the beach.

Left: The *caique 'Vierge de Lourdes'* at Fécamp, one of four built just after the Second World War by Jouen-Fiquet de Fécamp.

capstans until the 1860s, when they moved their boats to Honfleur, where they could stay afloat. Thus, the shape of the boat evolved into fuller hulls with a keel and a canoe stern. However, not rare in enclosed maritime communities, the name *platte* remained. The new boats, built exclusively at Honfleur, were between 9 and 10 metres and were originally rigged as *chaloupes* with two masts of equal length and with a topsail on the main (mizzen) mast. Some even had three masts with a small triangular mizzen, though over time the rig developed so that by the twentieth century they were chaloupe-rigged with a jib on a long bowsprit. These were regarded as manoeuvrable vessels, crewed by three men and a boy, and elegant craft. They had a sheer line painted in bright colours and an equally bright painted moustache. They generally trawled for sprats, shrimps, prawns and herring around the estuary of the River Seine, although some also worked westwards in Caen Bay, although they still landed into Le Havre.

The *picoteux* were small open boats found all along the coast of Calvados, part of Normandy between Le Havre and the Cherbourg peninsula, until about 1960. They were said to be extremely ancient, supported by their appearance, and were about thirteen feet long and capable of holding two or three men. They were pointed at both ends, had no keel and had a very rounded stem. Their name came from method of fishing for flatfish, called *picot* or *flondre* (literally 'spike'), and it was said that the name *picoteux* was commonplace back in the seventeenth century.

Although they were generally rowed by three crewmembers, each having two oars, they were also rigged with one dipping lug. Similar craft worked from other villages along this coast such as Courseulles-sur-Mer. Those few still working after the Second World War were fishing mussels in the Bay of the Veys as well as fishing for cockles, eels, bass, mackerel and flatfish.

Along the coast of the Cherbourg Peninsula the *vaquelotte de Cotentin* was a small fishing boat which fished inshore. Also called a 'boat Barfleur' or 'canoe Bourcet', it was the principal vessel working this coast between Cape de la Hague and the Bay of the Veys and had been since the nineteenth century. With a length of between 5 and 7 metres, they were rigged with the *Bourcet-malet* rig of large lug main, mizzen lug sheeted to a long bumpkin and jib on a long bowsprit. Once again the masts were set at the extremes of the boat, the main in the eyes and the mizzen against the transom. Regarded as strong and rustic vessels, the vaquelottes were used for all kinds of fishing: lines, fixed nets, drift-nets and potting. Several still sail, such as the 6.5-metre 'Seven Brothers', built in 1932 by Charles Bellot of Barfleur. It was motorised in 1947 and worked the fishery until retirement in 1990. Others include 'Jacques', built in 1937, also by Bellot; another Bellot vessel, 'Normandy', built in 1946; the oldest, 'Celine', built 1912; 'The Angelus', built in 1943 and now based in Omonville; 'Albatross', built in 1932; 'Sainte-Marie', built in 1948; and 'Louis-Simon', built at Barfleur in 1929. Two others are on display in the museum on the Island of Tatihou, just off the east coast of the Cherbourg peninsula.

Bautier de Barfleur was a smack-rigged counter-sterned vessel named after the 'baux', the long-line that it fished with. These craft were originally pole-masted, though after 1907 a higher mast allowed a topsail to be set, sometimes on a jackyard. They had a cabin with four berths and a stove. Woe betide any boy who, when emptying the embers of the stove, threw them to leeward when the lines were down. Fish then would not come!

Much has been written about the *Bisquines* found in the Bay of Mont St Michel and based in Cancale and Granville, and they are probably one of the best known of French traditional craft. Although the rig is based on the *Chasse marées* (literally 'tide-chasers'), the hull comes from the double-ended launches (*traineras* or *Biscaienne*, from which the word 'Bisquine' comes) used in the eighteenth century by the Basques. They originated as two-masted coasters, but the third mast was added for greater sailing balance. The straight stem and deep forefoot was efficient for fishing – especially long-lining. Later they were used for trawling and oyster dredging.

The early boats were found throughout the Normandy coast. They were undecked until about 1850, when they were generally decked over. Sizes increased through the habit of racing in annual regattas; the bigger the boat, the more successful. Two builders especially were renowned for their craft – Bouchard and Lhotelier – both working in the harbour of La Houle, Cancale, although the Julienne shipyard in Granville also contributed much to its development. They were carvel-built of oak on oak frames, with massive scantlings, typical of many Breton boats, and a counter stern, not typical. Towards the end of the nineteenth century the Bisquines were typically 14 metres overall, 3.5 metres in beam and had a 1.5-metre draft. However, they set a massive 100 square metres of canvas. They were easy to sail and work,

The *vaquelotte 'Henri-Josett'*, built at Barfleur in 1946 and now owned by the Maritime Museum of the island of Tatihou. (*Photo: Musée Maritime de l'île Tatihou*)

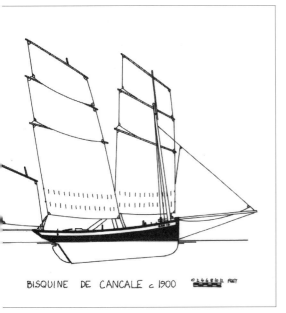

Above right: The *bisquine* is perhaps one of the best known French craft.

A *bisquine* sitting upon the beach at Granville, the depth of its hull apparent. Note the railway trucks upon the quay.

having no standing rigging, apart from running backstays, to hamper the fishing. The last one, '*La Perle*', was built in 1905 and she ended up abandoned on Epi Strand near Cancale around 1930. But for the foresight of one Frenchman, Jean Le Bot, who took the lines off the rotting hulk before it was too late, these boats would have been lost forever. Lifting the lines enabled replicas to be built, of which there are two – '*La Cancalaise*' (1987) and '*La Granvillaise*' (1990). The first was a copy (almost) of '*La Perle*' and the latter replicated the '*La Rose-Marie*', which was built at the Julienne yard in 1899. Today it is said that '*La Granvillaise*' is the livelier of the two, though they exist together in an atmosphere of complete harmony.

BRITTANY

Brittany is perhaps home to the best naval traditions that can be found in France. A ship crewed by Bretons was said to be a vessel that could sail anywhere, and during wartime was the most feared by the English fleets.

The Veneti of Brittany (based in Vannes and the Gulf of Morbihan) were one of the earliest boatbuilders, in business at the time of the Romans. Their boats were strongly built of oak with iron fastenings, with high stems and sterns and flat bottoms for beaching. They were built to withstand the heavy seas of Biscay. The earliest Roman *sinagots* were said to have huge iron fastenings – as thick as a man's thumb – and thin, supple animal-skin sails, similar to the Veneti boats. These boats were roughly 8 metres overall, and in Roman times were used for trading around the coast. It has been suggested that the Vikings, on their visits to the Gulf, took the idea of high stems and sterns back with them to copy, but subsequent discoveries seem to contradict this. What seems more likely is that the Vikings influenced the design of the *sinagots*: they became double-enders, and were used for fishing – mostly oyster dredging around the Gulf. More of these later.

The boats of the Veneti were generations later to influence the bigger Breton boats: the great *Chasse-marées* and the *lougres* (luggers). These three-masted boats were common in the eighteenth century, and during the two wars of 1740–64 these vessels, crewed by Bretons, became the terror of the seas for English boats – indeed, the English were so impressed with their performance that they copied the design into the three-masted fishing vessels of the south coast. When not at war, the French three-masters were used for servicing the sardine fleets off the Biscay coast, and, during the off season, for coasting.

By the turn of the twentieth century, two distinct types were to be found: the larger boats of 20 metres overall were used for coasting while smaller boats, being more manoeuvrable at 11–12 metres, were used for inshore fishing. Many of the bigger boats were used for smuggling across the Channel to England as they were extremely fast under their lug rig. Some even sailed to South Wales to pick up coal.

Originally three-masted, the middle mast disappeared late in the nineteenth century so that most became two-masters. Within a few years the third mast was re-adopted

SINAGOT DU GOLFE DU MORBIHAN c1920

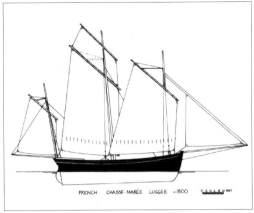

FRENCH CHASSE-MARÉE LUGGER c1800

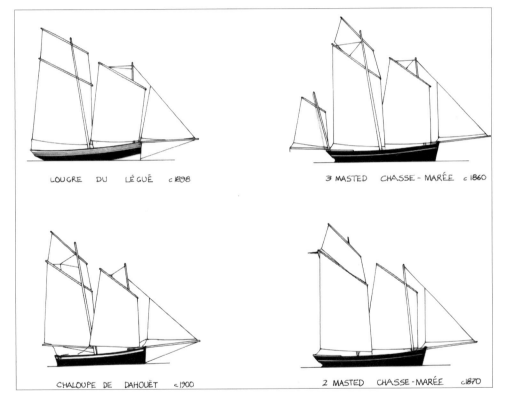

LOUGRE DU LÉGUÉ c1898

3 MASTED CHASSE-MARÉE c1860

CHALOUPE DE DAHOUËT c1900

2 MASTED CHASSE-MARÉE c1870

by some boats because they found that with only two masts the vessels suffered leeward drift. So popular were these strong, powerful boats that they were to be found all around the Channel ports (as we've seen) in the early part of the twentieth century. Their decline only came about when the motor was introduced.

The sardine is a member of the pelagic family of fish – the same family as herring, mackerel and pilchard, the latter simply being a mature sardine. Like other pelagic fish, sardines appeared in huge shoals all along the coast of Brittany from the beginning of April to the end of October. A huge industry grew up in the nineteenth century all along the coast. To supply this market, fleets of boats from both large harbours and small beach-based communities fished by day. They would set drift-nets in the early morning and lure sardines into the nets by sprinkling cod's roe into the sea. Huge amounts of this roe were imported from Norway in its salted form – called *la rogue* – hence the fishing became known as 'rogue fishing'. One of the earliest sardine boats appears to have been the *forban* of Bono, a town on the River Auray which flows into the Morbihan, the boat being a two-masted lugger similar to the *sinagot*.

Larger fish were caught out of season by setting drift-nets deeper. In Douarnenez, the biggest sardine port, some 4,000 fishermen and 700 boats were actively involved in the fishery towards the end of the nineteenth century. For this rogue fishing, a vessel was needed that could cope with some of the worst seas in Europe, was easily manoeuvrable around the net, and was capable of travelling long distances. The type that became popular was known as the *chaloupe sardinière* or sardine sloop and seems to have been an improved version of the *forban*. These boats were typically Breton in their shape, were double-ended, between 8 and 12 metres long and were renowned for the beauty of the sterns. One such example is the engineless replica '*La Barbinasse*', built in 1997.

In Breton tradition the sternpost of the transom was heavily raked – it has been suggested that the French introduced this into Scotland – and so the stern was rounded at deck while being pointed at the waterline, producing a beautifully-shaped body. These boats were completely open until the end of the nineteenth century, yet it was not unusual to find boats from, say, Douarnenez working from ports as far south as La Rochelle.

With their two raking masts with lugsails, and heavily raking sternposts, they were perhaps the most extreme type of Breton hull shape. Further south, such as at Saint-

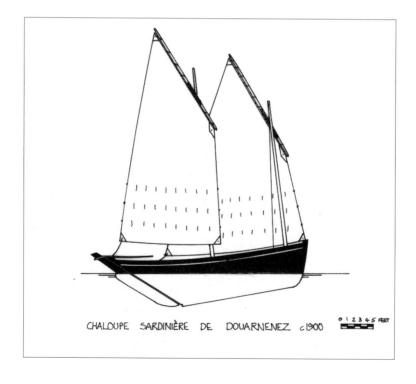

CHALOUPE SARDINIÈRE DE DOUARNENEZ *c*1900   0 1 2 3 4 5 FEET

Above: A *chaloupe sardinière*
dried out at Douarnenez,
showing its heavily raking
sternpost and sloping keel, a
reminder of some of the Scottish
boats of the late nineteenth
century (*Zulu*, Lochfyne skiff).

Right: A deck view of the same
vessel.

Gilles-Croix-de-Vie and Les Sables-d'Olonne, similar three-masted sardine *chaloupes* (at the former) or *canots* (at the latter) had transom sterns, an influence from southern France and northern Spain.

At the same time many other boat types were working off the Brittany coast. The *chaloupe de Dahouet* – a replica based on traditional lines has been built recently – again seems to have influences from the *Chasse-marée*, and local influences that arose from work in both pilotage and trawling. Similarly the *lougre du Légué*, a boat of 12–14.5 metres used for oyster dredging and beam trawling, had evolved from the *Chasse-marée*, although when shedding one mast it was the mizzen that was deemed unnecessary so that the rig is carried on a foremast set well forward and the mainmast amidships, the latter having a topsail. A large jib was set on the bowsprit, though by 1914 these boats were in decline. A replica vessel, '*Le Grand Léjon*', was built in 1992 and is well known along the Breton and Cornish coasts. The *langoustiers*, the lobster boats of the northern Brittany coast, have a similar hull shape, though they were gaffers with the typical cutter rig of main gaff, topsail, jib and staysail. This rig was more convenient for the stopping and starting routine of handling the pots than was the lug. Some of the *langoustiers* were well boats (*bateaux vivier*), with a central wet-well open to the sea to keep the lobsters alive.

The *Camaret langoustiers* were slightly larger vessels and these usually set a small mizzen right in the stern in improve manoeuvrability around pots. Lobsters were then passed onto suppliers, who were mainly around the Morbihan, and much of the trade to the European markets was in wet-welled boats such as the *dundée caboteur a vivier* or the larger *goélette a vivier*. In the first years of the twentieth century, larger *dundée langoustiers* sailed down the Spanish and Portuguese coasts, down the Moroccan and Mauritanian coasts or into the Mediterranean in search of new grounds. Some went north, exploring the Irish and Hebridean grounds, and a few even sailed as far as the West Indies. These ketch-rigged gaffers, with topsails and two foresails, appeared to be bigger versions of the counter-sterned, high bowed Biscay boats.

The *sloup sauzonnais* was a small boat confined mostly to Belle-Île in the late nineteenth century, developed so that the gaff-rigged boats could cope with the decline in the sardine fishery by following other fisheries. They simply set a large gaff main on a boom overhanging the stern and a large jib. Sloops also fished for mackerel – one example being the *sloup maquereautier du Croisic*.

Small boats also came in various types and sizes. The *canot a misainier* was a cheaper, smaller version of the bigger sardine boat which, as the sardine fishery declined, began to lose money. These could also be worked single-handedly. Their average length was 6 metres and they only had one mast, placed well forward with a standing lugsail – *misaine* meaning forward lugsail. Like the *chaloupes*, they were easily manoeuvrable because of their shallow forefoot, while the deep after-foot gave a good grip on the water. The raking sternpost was as steep as the bigger boats as they were taxed on the length of keel. Short keels also give the lowest wetted surface area, hence low resistance and thus greater speed.

Most, though, had transom sterns. They were carvel-built in good Breton tradition, using local larch on massive oak frames, although by Breton standards they were lightly

Camaret *langoustiers* at Newlyn, Cornwall in 1923. There's always been a good working relationship between the Breton and Cornish fishers. During the Second World War, many Breton boats sailed over to Newlyn and Penzance to escape the German occupation.

Above left: The 1992-built Camaret *langoustier* 'Belle Etoile' is a replica of the 1938 vessel of the same name. Some fifty of these vessels were built in the town.

Above right: A busy fishing scene in the natural harbour of Saint-Guénolé in about 1955. The boats here are mostly *langoustiers*.

CANOT A MISÀINE (MISAINIER) c 1910
0 1 2 3 4 5 FEET

The highly manoeuvrable *misainer*, so called for its *misaine*, the forward lugsail.

built. Having seen the scantlings of other, smaller boats, this is perhaps not an understatement, yet having sailed in Biscay in winter, it's not that surprising. The rig was extremely simple as there was no standing rigging. Mast rake was adjustable, the mast itself keel-stepped, wedged into position using a wooden wedge. Generally these boats carried no ballast, such was their weight. These, like the *chaloupes*, declined in numbers after the First World War. When one considers that at the peak of the sardine fishery some 60,000 people were involved in the industry – 25,000 of them fishermen – it follows that there must have been huge numbers of these two types of fishing boat. Some suggest there were 4,000 by the end of the century. Today, it's only a handful of survivors and replicas!

The *flambart* is another small boat from the north coast (some were also to be found in Normandy). This is a two-master, both masts of equal length, upon which are set two large lugsails, the after one being set on a boom that extends almost half its length over the transom stern, the latter earning the boat the nickname of 'ass-square'. They were used mostly for sardine fishing, oyster dredging and even transporting seaweed.

Another smaller craft of the north Brittany coast was the

Bourcet-malet rigged *canot jaguen*, a small open boat that worked long-lines on the inshore waters between Cap Fréhel and Ile de Cezembré in the Gulf of St Malo, the name coming from the inhabitants of the village of Saint-Jacut-de-la-Mer – the *jaguen*. The 13-foot '*Etoile du Matin*' is one such vessel, which was built in 1952 in Saint Malo and which has recently been restored. The sloop-rigged *dragous de Saint-Jacut* is chiefly a small trawler, the name coming from trawl gear (*la drague*) used. The latter also operated out of Saint-Jacut-de-la-Mer, where shelter could be gained when dried out around a small headland on the western side of the peninsula. A replica *dragous* was built and launched in about 1988.

The *sloup du bas Léon*, a small-boomed, solidly-built gaffer, between 5 and 7.5 metres, was used for potting and collecting mussels in the exposed waters around the north-west tip of Brittany, based at places such as Le Conquet, L'Aber-Wrach and Portsall.

Coquilliers were found in the Rade de Brest. These scallop dredgers also dredged for mussels and oysters, especially when the time they were able to fish scallops was rigidly controlled. They were originally lug-rigged, but became gaffers for ease of their work.

Small two-masted undecked sloops under 6 metres, called *bateau kerhor* after the town of Kerhuon, also fished within the Rade de Brest and its rivers. They also ventured outside the Rade into the Iroise Sea and fished the Chenal du Four and around the islands inside of Ushant, Béniguet, Quéménés and Lytiry. At these times they used Le Conquet as a secondary base, going home by train every fortnight. They fished for six days before returning with their catch. To sleep aboard the boat, they set an awning over three-quarters of the boat, bent over the lowered mast which sat in the crutch. Straw mattresses provided bedding and the four crew slept head to toe. Food was cooked on a small fire which sat on an iron plate balanced on the stone ballast. The fishermen were known for the spoons they made from scallop shells with wooden handles attached.

In the 1880s there were sixty *kerhors*, though numbers had reduced to forty by 1890. All were built at the yard of Hily in Pouldu, where they had been built since 1823, producing boats at the rate of a few each year until 1880. They also fitted the first motor in one in the early 1900s and by 1928 the remaining fleet were motorised. Before that they only used the standing lug sails when sailing downwind, preferring to row instead. Three thwarts crossed the boat and it is said that the boats from 1823 were almost identical to those of 1880, such was the success and perfection of the boat.

These boats fished with a seine-net, two men rowing the boat while the third fed the net out. The fourth man stayed ashore to hold the other end. The fish supplied the local area. Out of the four crew members, the boat owner was the skipper, who took a half share of the catch for himself, the boat and gear, while the other three shared the remaining half. Boats usually had 'Marie' in their name and at one time in the 1890s the entire fleet was named '*Marie-Jeanne*'! A replica boat – '*Mari-Lizig*' – was built in 1988.

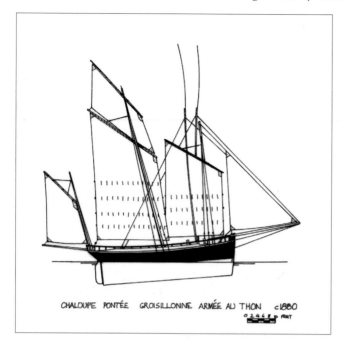

CHALOUPE PONTÉE GROISILLONNE ARMÉE AU THON   c1880

The other main fishery in the Bay of Biscay, apart from sardines, was tuna. Tunny boats – known as *thoniers* – were based in several ports towards the south of Brittany such as Les Sables-d'Olonne, Île d'Yeu, Etel, Groix (said to be the capital of tuna fishers), Concarneau and, later, Douarnenez. These were all either sloops or ketches (*dundée thoniers*), gaff-rigged with huge flaring bows, counter or transom sterns, and a hull shape similar to the other boats: sloping keels and raked sternposts. Their shape is indeed more reminiscent of Spanish boats designed for the longer Atlantic seas. Bigger *dundées* also worked out of La Rochelle to the south. Similar shapes are found in the lobster boats of Camaret. The only exception to the gaff rig seems to be the *chaloupe Groisillone*, a three-masted tunny boat with a lug rig descended from the Chase-marée boats.

The *sinago*, as mentioned before, originated in the Morbihan, like the *forban*. What became of the *sinagots* between the Viking times and the eighteenth century is vague, but by the early nineteenth century it seems that they were common around the Gulf, recognisable by their horizontally-peaked lugsails. They were of a very basic design with a long, shallow keel, a vertical sternpost, a locker below a small foredeck, and two benches: one to support the mainmast and the other for the crew.

Towards the end of the nineteenth century the lugsails obtained a higher peak as other fishing boats made their presence felt. They also grew in size and some had sloping sternposts. This generally gave them better working space. By the First World War, *sinagots* were commonly seen in places outside the Gulf such as Belle Île, Quiberon, the small islands of Houat and Hoëdic, and inside a line between the south of Belle Île and the Pointe du Croisic. They mostly fished for shellfish.

The *canot basse-indrais* was a small river boat from the Loire, the 'basse-indre' being part of the River Loire downstream of Nantes. They worked the river and out into the estuary.

The *Chaloupe de la baie de Bourgneuf* is a lug-rigged sardine boat unique to Pornic and Noirmoutier. These boats were double-enders, fully decked with a cabin, up to 9 metres in length and were chiefly used for trawling in the bay. The last vessel was built in 1908 so that by 1950 they had all disappeared until a replica vessel, the *'Jeanne J'*, was built in 2008.

Right: 'Nebuleuse', a *Camaret thonier* built in 1949 at Le Hir et Péron.

Below: Sinagots with very square-headed lugs.

A *chaloupe de la baie de Bourgneuf*.

VENDÉE TO BASQUE COUNTRY

Although Les Sables-d'Olonne was the first French town to send boats to the Newfoundland cod fishing in the late seventeenth century and the river was furnished with a pier and a wharf in 1768 and 1787, it wasn't until the mid-nineteenth century that the port was developed to any extent to allow the growth of the tuna and sardine fishery. This in turn led to the birth of the canning industry, once so important in the town.

As well as the three-masted sardine boats discussed above, the brightly coloured *gazelle des Sables-d'Olonne* was a beautifully shaped sloop – hence its name, from the swift and graceful antelope – which gained favour after 1906. Between 1880 and 1905 the sardine fishery was at its height in the Vendée, after which a decline set in, forcing the fishermen to go further offshore, and thus bigger boats were needed as their small craft were deemed unsuitable.

With Breton boats having fished alongside them, the Vendée fishermen were encouraged to buy in boats from ports such as Douarnenez and Concarneau. These they altered somewhat, raising up the bulwarks and changing to a sloop rig. But, by 1906, the local boatbuilders had begun to build their own versions, which were slightly less fine in entry and had a pointed stern and high bulwarks, though these were open at the stern beneath a hefty rail.

Thus the *gazelle* was born, a boat capable of dredging in winter and chasing tuna. However, they were short lived because, after the First World War, with developments in fishing techniques and the motorisation (which didn't suit these vessels) of other

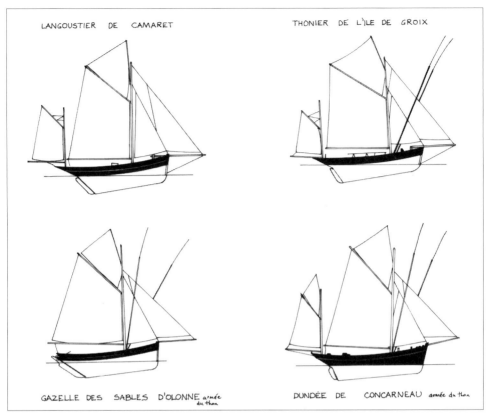

LANGOUSTIER DE CAMARET

THONIER DE L'ÎLE DE GROIX

GAZELLE DES SABLES D'OLONNE *armée du thon*

DUNDÉE DE CONCARNEAU *armée du thon*

A *thonier* from Les Sables d'Olonne in 2008. '*Vieux Copain*' was built in 1940 to work from Île d'Yeu. After being decommissioned in 1974, she worked in Greenland until returning to Granville, where she sank. She was raised and taken to Paimpol, where she was restored to her present state.

sardine boats, no further *gazelles* were built after the early 1920s. Today, a class of small sailing craft at Les Sables-d'Olonne, developed mainly for racing, has been modelled on the larger craft.

The *coureauleur* is a type of vessel the specific name for which comes from '*coureau*', which is a sea channel separating a shallow island close to the mainland, such as that of Île de Ré and Île d'Oleron. Thus they were fishing the shallow expanse of water between the two islands and that between the north-east coast of Île de Re and the mainland – known respecively in French as *le pertuis d'Antioche* and *le pertuis Breton*. Both these areas of sea face the Atlantic, where short, sharp seas can be extremely dangerous to any small boat. However, they were rich in fish.

These were small cutter-rigged vessels of 7–9 metres, often referred to as sloops, perhaps using the older definition of a sloop as being a rig with a single mast located forward of 50 per cent of the length of the sail plan. In this traditional definition a sloop could have multiple foresails, as in this case. They were shallow draft vessels with a full form and flat floors, enabling them to be beached. They also had a good carrying capacity, useful when loading oysters, as they were generally used for dredging oysters and scallops, trawling, potting and growing mussels. They are not to be confused with the barges of the Gironde and Dordogne, which are also called 'coureaux'. One surviving coureauleur is the 1909-built '*L'Argo*' which, with a length of about 7.5 metres, has a draft of one metre.

The *Filadiere de Gironde* is a type of small river boat used for fishing in the Gironde. A *lasse* is a small, flat-bottomed punt once used to carry oysters.

The *Pinasses d'Arcachon* were small fishing boats developed in the nineteenth century from the *pinasses* of the Gironde, which were primarily small sardine boats built of pine which themselves had evolved from the 'watcher' boats that had sailed out searching for the sardine shoals prior to the main fleet setting out. They were lightly constructed, hence they had a good turn of speed, and their name is said to have come from the English 'pinnace', which they resemble. They fished inshore, being found all along the coast from the Morbihan to the Gironde and further south. The Basin of Arcachon, being an inland sea, was perfect for them and they were soon used for oyster cultivation following its expansion after 1860. However, they were superseded by the *bac a voiles d'Arcachon*, barges which were more suited to transporting oysters and the roof tiles used to grown the spat on than the *pinasses*.

Finally there are the *trainieres* and *chaloupe basques (txalupa)* and their earlier version, referred to as *txalupa handi* – the big rowboat. Whereas the bigger boats were the traditional seine-net boats of the Basques, the *txalupa handi* boats, which also had two lugs, were used for all manner of work, including far-off cod fishing, towing dead whales to the shore and inshore fishing closer to home. Some were even converted as pirate boats by the addition of a cannon between the masts. They continued in use until about 1925 and a replica, '*Brokoa*' – meaning 'Gannet' – was built by the Hiruak Bat Shipyard in Socoa on the Bay of Saint-Jean-de-Luz in 1991. These boats will be discussed in more detail in the next chapter for, as everyone knows, the Basque territory overlaps the border between France and Spain.

A *pinasse* from Arcachon in the
late nineteenth century.

SOURCES

France has a whole host of publications regarding maritime history and plenty of
these concentrate on the various fisheries. There's also the *Chasse-Marée* magazine,
with articles on working craft. They also have a website through which books are
available. One that gives an overall picture is *Guide des Voiles de Peche*, from *Chasse-
Marée*. However, this and the others are all in French. Two English books consulted
are: *Mariners of Brittany* by Peter Anson (1931) and John Bethnell's *Boats and Sails of
Morbihan* (1991). One book of great interest concentrated on the *flobarts* is Bertrand
Louf and Francois Guennoc's *Les Flobarts de la Cote d'Opale*.

The fishery museums visited were at Dieppe and Concarneau, though there are
more general museums that can be found on the internet, such as the one at Tatihou
Island, which have a small collection of historical boats. The Brittany coast is simply
soaked in maritime history and the festivals at Brest (every 4 years) and Douarnenez
(biennial) are always a showcase for French traditional craft. The *Bassin d'Arcachon*
is worth a visit just to see how oyster spat is grown on roof tiles!

CHAPTER 12

The Atlantic Coasts of Spain

The Gulf of Gascoign to Gibraltar,
Excluding Portugal

THE NORTH AND NORTH-WEST COAST

The coastline of Spain spans some 3,100 miles, over half of which borders the Atlantic Ocean while the other half lies within the Mediterranean (see chapter 15). This Atlantic coast is rocky, especially around the north-west corner where the notorious 'Costa da Morte' is the graveyard of many a ship. The short west coast is indented with inlets protected by islands while the much longer north coast borders the Bay of Biscay and is a diverse string of sandy beaches, cliffs tumbling into the sea and sheltered harbours and fishing ports.

The Spanish consume three times as much fish per head as the average European, more than anyone else except the Portuguese. Consequently, they have to keep what they catch for their own discerning population. They have the second biggest fishing

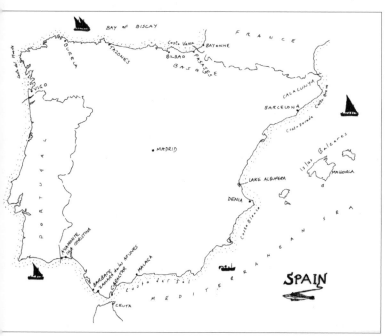

fleet, with about 17,500 fishing boats employing 90,000 fishermen in 2000 and many more shore-side. They also seem to have the worst reputation, although as is so often the case, headline instances are grossly exaggerated. The majority of Spanish fishermen are happy to fish in their coastal waters, serving the local fish markets with daily supplies of fresh fish.

Until a few years ago, it was estimated that, out of the 17,500 fishing boats,

15,000 fish inshore. Most of these worked off the north coast, itself said to consume more than three times as much fish as the rest of Spain. Indeed, Galicia was home to the largest fishing fleet in the world in the 1950s. Only a small percentage of these work from deep-sea harbours such as Vigo or Burela, the former being the largest fish port in the country.

To follow the evolutionary process that brought about the modern trawlers in the Spanish fleet, we must first go back several centuries to the early pioneering days of fishing. Although there are perhaps two distinct and separate areas from a fishing point of view in Spain – the Atlantic and Mediterranean coasts – this chapter only covers the vessels of the former.

The Basque region – Euskadi – stretches from Bayonna, in France, right along the precipitous coast to Bilbao. The language, Euskera, is about the oldest in Europe and its roots are unclear. It's not surprising, then, that the Basque folk are fiercely independent and were clearly among the earliest settlers in the Iberian Peninsula. Consequently, they were fishing long before anyone else; long, long before Bravo Britannia began to rule the waves.

The earliest records of Basque fishing start about AD 600, although they were at it well before that. Prehistoric Basque-man was at it, judging by drawings of fish found in a cave at Guipuzcoa in 1969 thought to be 12,000 years old. So was Roman Basque-man, who introduced salting on a large scale. Cliff-top watchers alerted the crews of their small open boats, much like the Cornish pilchard huers, although it has to be said that they regarded the sea as an unknown quantity, so they didn't venture too far out, especially as the whaling season began in November. The Bay of Biscay remains today as much of an unpredictable environment as it did then, especially in winter.

The Vikings came to the Basque coast in the ninth century and the Basques quickly adopted the idea of clinker boat construction. Their vessels became more seaworthy and able to sail longer distances in search of a new fish called cod. Like whaling, the smaller boats carried on their ships were used to catch the cod. The North American cod dories evolved from these craft. The Basques soon became the principal suppliers of both whale meat (approved by the Church for consumption on 'meat-free' days) and cod throughout Europe, and the best boatbuilders. Salt came, almost exclusively, from Portugal.

Nearer to home, outside the whaling season, the Basques fished for bonito, hake (*merluza*), sardines and anchovies. However, sea-bream (*bixugu* in Euskara or *besugo* in Spanish) soon became their principal catch, and the winter, too, was the prime fishing time for this. Traditionally, the Spanish eat sea-bream at Christmas, although supplies are so short nowadays that this is rare. Sadly, the farmed variety is spreading all too quickly.

Most of the fishing was done by seining – the spreading out of a net in a circle around a shoal, much in the same way as the Cornish caught their pilchards – although hand-lining and tuna-fishing techniques improved. Hand lines were dropped over the side, with each of the four crew controlling four lines, while the skipper, from his

Setting the seine-net around a shoal of sardines in the Basque Country.

seat at the stern, only had one. For catching tuna, cacea or *currican* lines were used. These consisted of long poles, with three lines from each pole being suspended over the side.

Another reason that fishing developed in this part of Spain first was the closeness to the rich fishing grounds that lie on the continental shelf. This shelf is only some 40–50 miles offshore from the Basque coast, whereas further west boats have to travel twice that distance. The coast of Galicia is also far more exposed, as anyone sailing off Finisterre will certify.

The boats these fishermen used came in three distinct sizes and were often referred to as *txalupa*. The smallest, the *batel*, was under 6 metres long, and rigged with one small dipping lug. These boats hardly sailed more than a mile offshore. The next size up was the *potines*, usually about 7–8 metres long, while the largest, the *bonitera*, obtained its name from its tendency to chase after the bonito, a fish like a tuna. These were up to 12 metres long and were entirely open. They were double-enders, narrow for their length, and had up to eight thwarts to carry up to fifteen crew. The *txalupa handi* (big row boats) were perhaps the earlier versions of these. One model of a *bonitera* seen in Bermeo had leeboards, so shallow were these craft. However, it must be noted that this was the only instance of a leeboard, as they were rarely used. The *bateles* (plural) were rigged with either one or two dipping lugsails, the earliest of the two-masted versions having a small foresail in comparison to the amidships-mounted mainmast, while later boats had the foresail as the biggest. One early characteristic of these was the rounded bulbous stern, not unlike the cruiser-stern of the Scottish MFV, although latterly, after motorisation, this became commonplace. Prior to this, many had sharp sterns, off which the rudder was hung. Like all the contemporary craft of Spain, they were built locally by the master, the local shipwright, who was always a

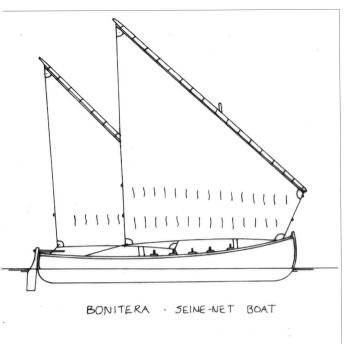

BONITERA - SEINE-NET BOAT

Above right: A typical two-masted *bonitera, c.* 1870.

respected member of the community. Boats were largely built entirely of oak, although pine was sometimes used for planking.

A *batel* was built by the Centre of Research and Construction of Traditional Boats at Ontziola, a traditional boatyard in Pasaia Donibane for the Pasaia, Euskal Odisea – Pasaia, the Odyssey of the Basques. Rigged with two dipping lugsails, the boat is about 9 metres long but appears to be nearly as fast as the larger boats. The largest were reported to have sailed at twelve knots to be the first out to the sardine shoal.

Use of these seine-net boats spread right along the north coast of Spain and round the corner to Vigo. Indeed, so impressed were the Portuguese that they introduced them into their fleets around the turn of the twentieth century. Literally hundreds, probably thousands, of these craft worked offshore in the nineteenth century.

The boats of Vigo differed from their Basque counterparts in that they rigged lateen sails, the furthest north this tendency is said to have spread. Although supposedly only a Mediterranean rig, its use was common right along the western coast of Portugal. The seine-boats of Galicia were said to have acutely sloping stems while those of the Basques were upright.

That the Spanish were fishing in the sixteenth century is confirmed by reports that Francis Drake went south to hinder plans by Philip of Spain in his preparations to send the Spanish Armada to invade Britain in 1588. Drake, it is said, seized several fishing boats and nets, hoping to deprive the Spanish of supplies for their warships. The boats, according to one report, were rigged in much the same way as they were at the beginning of the twentieth century.

A Vigo-type seine-boat used to catch sardines.

In the fifteenth century the Basques were fishing off Iceland and the Faroes, sailing there in the big caravels and fishing for herring and hake with the smaller seine-boats. The same fishermen had already begun to exploit the lucrative Newfoundland cod fishery, it being said that they arrived off the North American coast at least a hundred years before Columbus and Cabot. This was the fishery that was to survive for centuries until its pillaging and eventual closure in the early 1990s and Spain, although not the only one, had to accept a share in the responsibility for the overfishing of this cod.

The coast of Galicia has long been a graveyard of ships. From the northern fishing harbour of Camarinas southwards to Cape Finistera, it has the daunting title of the Costa da Morte – Coast of Death. However, numerous fishing communities grew up at this, the very edge of the sea, as fishing boats grew larger and more deft at venturing further offshore. Nevertheless, the port of Pasajes, in Euskadi, its harbour being built in the nineteenth century, remained the premier Spanish port for fish until the early years of the twentieth century, with landings of cod and sea-bream the highest.

Steam was introduced into the Pasajes fleet on a large scale in the 1920s. However, the roots of steam go back another thirty years or so, although Blasco de Garay, a Basque engineer, produced a plan for a ship powered by vaporized water as far back as 1543. King Carlos I was not impressed, so the idea was shelved.

After a gale in the late 1870s, when over 200 Basque fishermen were drowned, a steamer was built in San Sebastian to tow the seine-boats offshore. Within another ten years this boat was being used to trawl and the arguments so well documented in Britain about trawling began. Meanwhile, the first steam-seiner was built in the

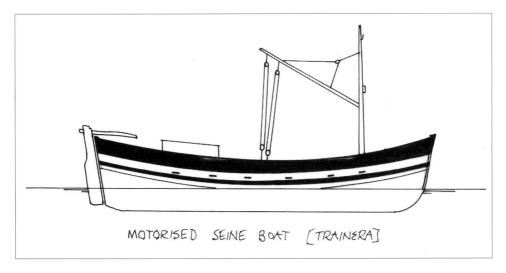

MOTORISED SEINE BOAT [TRAINERA]

French Basque port of Saint-Jean-de-Luz. The near continental shelf waters soon became exhausted so that the coastal Basque fishermen had to search wider in distant waters to fill their nets. Then, with an improving railway infrastructure and because of its proximity to richer fishing grounds, emphasis in fishing moved westwards, so that Vigo eventually became the prime fishing harbour of Spain and Galicia the home to the largest fleet in the world by 1950 in terms of landings.

Steam, like in the rest of Europe, played its part in changing the face of Spanish fishing. However, it wasn't until the advent of the internal combustion engine and the ensuing rush to motorisation that change came to the inshore fishermen. Small engine units were fitted into the bigger seine-boats, the *boniteras*, in the 1920s, and within a few years almost the entire fleet changed over.

Today the harbours of the north coast of Spain are still full of fishing boats, most concentrating on inshore fishing for shellfish, octopus, sardines, conger, eel and hake. Many of the smaller ones of these still resemble the old seine-boats. They've retained the cruiser-type stern yet have incorporated the huge rounded, flaring bows of the deeper sea boats. The latter have also become much deeper for extended ocean fishing, have the round, but fuller, bilges, and have retained the heavy rounded stern as well as the enormous bows to deal with the huge Atlantic waves. Amazingly, wooden boatbuilding is not a thing of the past, and many Spanish fishermen of the near-distant type insist on wooden craft, whereas steel boatbuilding, although growing rapidly, is still in a minority. Galicia still had thirteen shipyards working in wood and Euskadi one in 2002.

In specific areas, the design of the small motorised seine-boats has altered for a localised fishing. In Tazones, in Asturias, the open boats have become extremely full with rounded, slightly flaring, bows and round sterns, fitted with large-powered motors for lobster potting. Although the shape retains much of its ancestry, these 6-metre-long craft have evolved purely from the necessity of having to haul pots over the side. Engines have to be powerful to enable pot haulers to be added, much like Cromer crab boats and Cornish potters have developed over the years.

Another distinctly different type of craft still remains working from the north-west harbours of the country. These flat-bottomed, double-ended, lug-rigged *Canoas of Muros* (also known as *bucetas*) are about 5 metres long in the largest cases, and are built off a bottom that is made up of transverse planking fixed to a central keel, with a small degree of rocker built in. Three planks are then fitted either side, with framing, a transom stern and thwarts to strengthen. They were used for all kind of work, from transporting animals to fishing close to the shore, mussel gathering and even barnacle collecting from the offshore rocky islands. Being flat-bottomed, they were

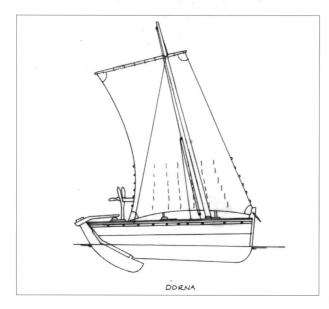

DORNA

both cheap to build and easy to haul ashore among the harbourless coastal communities. Furthermore, they are said to be good boats to sail in the tidal zones.

Another particular type found on the Costa da Morte is the *dorna* of the Ria de Arousa. These single lug-sailed craft bear similarities to the Breton *canot de misaine* and were used for lining, netting and potting. They were distinctive in the shape of their sternpost, had a small transom and came in varying sizes from about 4.5–6 metres. Similar craft, perhaps with wider sterns and albeit often aluminium, steel or fibreglass build, still work from the famous lobster harbour of A Guarda, close to Portugal's border.

The *lancha* was developed for the sardine fishery, working close to the shore – there was never much need to venture far out to sea – on the grounds around the north-west of Spain, considered as being among the richest in Europe until

A *dorna* sailing fast under lateen sail.

Lanchas in Muros harbour in 1947. (*Photo: Museo do Mar de Galicia*)

Hauling in the seine net aboard a motorised *racú*. (*Photo: Museo do Mar de Galicia*)

A motorised *batel* called *racú*. This boat came from the Cantabrian Sea and is similar to that called a *merlucera*. (*Photo: Museo do Mar de Galicia*)

recent times. They were double-ended craft with sharp, raked stems and sternposts and were vessels said to be almost symmetrical. According to Morling, when motorisation came, many owners converted the stern to the bow and had the original bow altered to receive a stern tube. The average length of a *lancha* was around 7 metres, and most had fore and side decks, although the area of these was increased after motorisation. They were often called a *lancha de relinga*, which referred to the particular sail – *vela de relinga* – a form of the lugsail that is set on a short yard on a mast that has an adjustable raking which lowers the centre of the sail area, thus improving performance. Although the *lancha* survived well into the twentieth century as a workboat, it was regarded as a heavy boat and, although smaller and lighter and built in fibreglass, the *buceta* of the Galician coast has similar lines to the older *lancha* and today many have been motorised.

The final Galician boat found along the west coast is the *racú*, which were up to 30 feet in length. They developed from the earlier *batel* and were used mostly for fishing the Cantabrian Sea and coast from the beginning of the twentieth century until around 1970, when they declined. They were initially rigged with one simple sail similar to that of the *dorna* until engines were fitted in the early 1960s, after which they became to be known as *Motoras de Bouzas*, Bouzas being near Vigo. Similar to the *merlucera*, their stern developed into a rounded and wide one from the earlier fine after end, and the hull had to be strengthened.

THE SOUTHERN COAST

Moving away from the rich fishing area of the north, the south Atlantic coast of Spain is short, yet several fishing harbours are still active between Ayamonte on the River Guadiana, the south-western border with Portugal, and Tarifa, Europe's most southerly point. Here, tuna fishing was traditionally the main preoccupation of the fishermen and lateen-rigged *caiques*, similar to the Portuguese tunny boats (to be seen in the next chapter), worked off the exposed coast, normally from the shelter of the various rivers. Other boats were the open canoe-types called *galeas*, which fished for sardines and were said to have been crewed by upwards of forty men. One exception is what is now the resort of Zahara de los Atunes, where 5-metre-long transom-sterned boats still work from the beach. These have developed from the earlier *caiques*, adopted for outboard motors with the transom. A few miles west, at the small town of Barbate, at the entrance to a small river, traditional double-ended tuna boats still survive. These long shallow boats, all painted a greyish blue, are similar to the northern seine-boats, and are rowed out in the summer to catch the tuna close inshore. Several which were, until recently anyway, still used for today's tuna fishery, can be seen sitting on the opposite shore to the village in the off-season. It is assumed that these have evolved from the half-decked boats which operated beach-seines right along into the Algarve region of Portugal.

Today's modern ubiquitous fleet that resembles that of most of the Portuguese coast and into the Mediterranean, with their transoms and high flaring bows to cope with the large seas the Atlantic can throw at them, is hardly worth a mention.

SOURCES

Much of the information came from a circumnavigation by the author of the entire coast in 1999 (by road). Bermeo Fishing Museum has some excellent material. The Museo do Mar de Galicia at Vigo has many collections but unfortunately was not visited at that time. It has since been refurbished and has been helpful with the preparation of this chapter. Staffan Morling's book *Lanchas and Dornas* (Batdokgruppen, 2003) is the most wonderful study of regional craft and is written in English. Otherwise, there is a whole host of good books on these craft, all written in either Spanish or Euskera.

CHAPTER 13

Portugal

The River Mino to the River Guadiana

For its size, Portugal has more fishermen than any other European Union country. In comparison to its larger neighbour Spain, which has a population of nearly 40 million and 90,000 fishermen, Portugal has a quarter of the population and almost 40 per cent of the number of fishermen (34,000). It also has one of the finest and most varied assortments of fishing craft in the world, craft that sail out into the Atlantic to catch millions of tons of any number of different species of fish to feed a population that eats the highest amount of fish per head of any European country. This is largely due to the fishing grounds that lie close to its shore, enriched as they are by the close proximity to the continental shelf, but it equally might have something to do with the fact that the country is one of the oldest nation states in Europe.

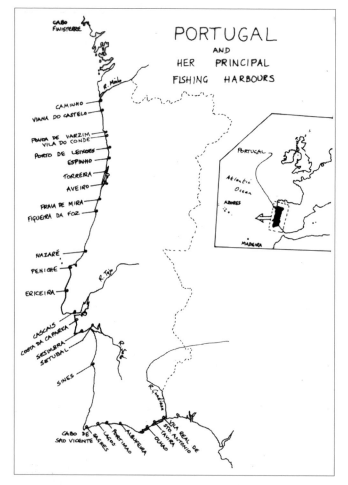

The coastline is over 500 miles long and can be separated into three distinct areas - the north coast that stretches from the Spanish border at the River Minho to Ericeira, the central area around the River Tejo at Lisbon, and the southern Algarve region which stretches from Cabo de Sao Vicente (Cape St Vincent) to the border with Spain at the River Guadiana. Each has its own types of craft that have evolved through the individual cultural backgrounds, although there are definite similarities between areas. The lug, lateen and sprit rigs all appear, and often the tendency is to paint the boats in bright colours to enable them to be seen at sea. The craft are mostly flat-bottomed to allow them to be easily beached. However, the first chronicled fishing craft is the *barca pescareza*, a fishing caravel mentioned as being sent out by Henry the Navigator during his first expeditions to Africa in the latter half of the fifteenth century.

The north coast is most definitely the most exposed. Atlantic gales often wash the shores with their savage vengeance, wreaking havoc with everything in their path. Harbours were sparse until the nineteenth century, and even these were often inaccessible when waves and surf made their entry impossible, given that they were nearly always close to the various river mouths that feed into the ocean. Beaches were the only other possibility.

One of the best known of Portuguese craft is the open *saveiro*, sometimes called the *meia-lua* off the southern beach of Costa da Caparica, where they are also common. This means 'half-moon', which aptly describes its shape. These craft worked off the exposed western beaches all along from Espinho in the north to Nazare further south, as well as at Caparica. They were flat-bottomed craft, over 11 metres long, canoe-like without a keel, that were launched directly into the surf using huge sweeps, up to 9 metres long, each of which were handled by up to five crew. They had really high prows to enable them to cope with the large waves of the surf. With four such sweeps being swung out on the largest craft – some smaller *saveiros* only had two sweeps – that means there were over twenty crew aboard, and another half-dozen men helped from the shore by pushing the boat out using another long pole against the sternpost of the boat, which, incidentally, had no rudder attached. Often a pair of bullocks was used for this latter task. Once out at sea, the boat was always rowed, no sails being carried. The boats themselves were used to set seine nets a mile offshore to catch sardines, the bridle-ropes of which were then brought back to the beach, and which were then hauled in onto the beach, again using either manpower or bullock-muscle. Up to twelve animals were used, taking turns, to haul in the huge net. They were also used for ground trawling with a crew of forty-two men according to Filgueras. Once

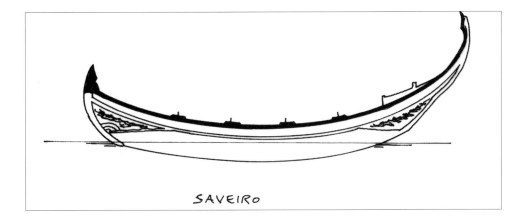

SAVEIRO

Launching a *saveiro*
off the beach at Praia
de Mira into the
Atlantic surf can be
quite an operation.
The boat itself has
at least six men
per oar (four oars)
while another dozen
push off with a long
wooden pole.

the boat returned to the shore, they were carried up the beach and they could be
turned around by rocking the boat longitudinally and swinging it round. Traditionally,
the fishermen's wives, the *varinas*, sold the catch ashore. Indeed, this practise still
exists, both on the markets and from improvised stalls set up around villages. Carts,
drawn by tired-looking horses, still hawk the day's catch inland of the Ria de Aveiro.
Today this means of fishing has largely disappeared, although there are still some
saveiros on the beach at Praia de Mira.

Nazare, once home to a beach-based fleet numbering over 200 boats, is also
renowned as being one of the last places where the fishermen worked directly off
the beach, although it has to be said that most now work out of the new harbour
built half a mile to the south of their beach. Here their motorised boats can remain
afloat throughout the year, although winter can severely restrict their working days.
Saveiros worked here, as did smaller open rowboats such as the *barca das Armações*
or *Xavega*, both of which operated nets close inshore. The larger *lancha da Galeao*
were lateen-rigged double-enders, half-deckers that worked a peculiar mode of
fishing called *cerco americano*. Although the boats of today hardly resemble those of
yesteryear, the fishermen can still often be seen dressed in the way they once were, in
their checked shirts and black stocking caps, mending their nets on the beach. The
fishwives also, with their flowing multi-layered petticoats, can often be seen drying fish
on wire racks on the beach. Dried salted cod (*bacalhau*), it's worth mentioning, is still
eaten extensively in Portugal, where it is known as *o fiel amigo*, or the faithful friend,
such is the reliance upon it when fresh supplies are scarce. Traditionally there are 365
different ways to cook it, one for each day of the year, although *bacalhau al forno* is
still *the* dish of the country.

About eighty miles north lies the village of Torreira, where, on its seaward side,
bullocks occasionally still help out with the local fishing. Catches, though, tend to be
poor, and the work is often carried out only for the benefit of the watching holiday-

A few *saveiros* do still exist; this one is seen in 2012 on the beach at Praia de Mira, where boats such as these are still used for fishing.

Saveiros on the beach at Nazare. See the bullocks on the right.

makers. Such is fishing nowadays in a country affected by massive imports of fish from its nearest neighbour.

On the other side of Torreira the town faces directly onto the Ria de Aveiro, the home of the seaweed-gathering *moliceiros*. These are canoe-type, shallow-drafted craft that have delicately painted, often humourous, decorations on their high prow. Gathering the seaweed to fertilise the land was obviously once a furious industry, there being 1,342 boats registered in 1889. They also gathered the *pilada*, the small crab they dried and then used as fertilizer on their potatoes. These boats were built by some three dozen boatbuilders who circled the lagoon. Today, only a handful of *moliceiros* sail the waters in the summer, setting their characteristic square lugsail on the centrally mounted mast, although similar shaped but smaller canoes still fish in and around the lagoon.

The practise of using bullocks to pull the beach boats around was common all along the north coast. At Ericeira they were used even though there is a harbour to keep boats at shelter afloat. On one particularly windy day, when the thirty or so open seine-boats were pulled well up the ramp, well away from the gigantic waves that were sweeping over the protective harbour wall, it was so rough that half the village's population was out gazing at the impressive sight of the ocean rollers. The boats, although now fitted with powerful engines in place of a rig, and apart from being fuller, still retain some resemblance to the traditional lateen-rigged boats, the *barcas*, *batels* (also called *Canotes*) and *focinheiras* that once worked the sardine seines and lobster pots off the beach.

North of Oporto, the *lancha* is a type of double-ended boat used widely, especially from Povoa de Varzim, where it becomes known as the *lancha poveira*. These are sometimes referred to as *catraia*, especially when they go in search of the *pilada*. A replica *catraia*, 23 feet long, was built in 1993 at Esposende, near Povoa de Varzim. The latter was one of the principle fishing ports of the country, around which there were said to be over 10,000 boats with a combined tonnage of 35,000 tons, i.e. each boat was small at 3.5 tons. Influence for all the northern craft is said to come from Galicia. Various versions of *lancha* worked different fisheries, such as the *lancha grande* that fished for whiting, while the *lancha pequena* caught sardines and other variants with long-lines. The former are up to 8 metres long, while the smaller ones are usually 7.5–9 metres. They are rigged with either a lateen sail or a dipping lug, often very squarely cut, and have four or five oars on either side. One 10-metre lancha, the first in recent years, was built at Povoa de Varzim in the early 1990s on the orders of the council of the regional museum, and is today crewed by up to thirty-two men. These *lancha* were suited to motorisation in the twentieth century and have survived all along the coast in a similar yet fuller shape, while others have transom sterns.

At Peniche, another *lancha* is used for the sardine fishing. It's worthwhile noting that 250 million tons of sardines are annually landed in Portugal, a high percentage going to the various canning factories that line the coast. These lateen-rigged craft were transom sterned and had a crew of up to fourteen men. However, Peniche is also famous as being where the Spanish seine-boat was first introduced into Portugal. This, in 1913, was to increase the yield of the sardine fishery, and led to more

Moliceiros on the Ria de Aveiro in 1999. These boats, used for collecting seaweed, can be delicately decorated.

Small canoe-type boats with outboard motors fishing in the Ria de Aveiro in the rain in 1999.

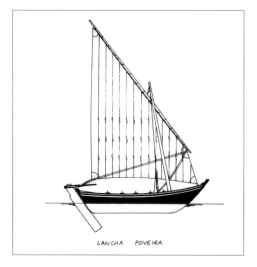

LANCHA   POVEIRA

than a dozen seine-boats, locally called *traineiras*, being imported that year. The following year one particular boatbuilder of Peniche began building his own version in what became known as the '*Vigo type of traineira*'. This was a 12-metre double-ended galley, rigged with one or two dipping lugsails, although lateen sails were later also used. They carried up to fourteen crew and had up to six sweeps either side. Like pilchard seining in Cornwall, these boats set a net around a shoal of sardines. First, once they'd detected a shoal using their knowledge of the natural appearances (*geitos*), much in the same way as the Scottish fishermen searched for herring, they concentrated the shoal by spreading cod roe as bait like the French sardine fishers did (the Spanish used bran). Then the net was cast with a buoy on one end, and the boat rowed around the shoal, pulling the net, until they could join up both ends. Acting like a purse-seine, the net was tightened to entrap the shoal, and then the complete net was rowed into shallow water for emptying into the boats. Sometimes it was pulled ashore and emptied directly onto the beach. Subsequent to motorisation, these *traineras* developed high flaring bows, long sweeping counter sterns, powerful engines and lengths of up to 17 metres.

Steam trawlers were introduced towards the end of the nineteenth century, and the open beach boats were converted to motor, firstly in 1924, although three years later nearly the entire Peniche fleet had been converted. The hull shape drastically altered, becoming larger and decked, and the new type became known as the '*Peniche-type of traineira*'. However, with the growth of the sardine industry, especially in the canning sector, and the introduction of the diesel engine, the seine boats became much bigger and evolved from the steam or 'fire' seiners into the deep-sea seiners with their haulers and gantries that replace the muscles and arms of the multitudinous fishermen. Today, the Port of Leixoes, close to Oporto, is Portugal's premier fish port, and is home to many of these large vessels.

The second distinct area lies around the mouth of the rivers Tejo and Sado, where, in places such as Cascais, Trafaria, Costa da Caparica, Sesimbra, Setubal and on the Troia Peninsula, various types of both deep-sea and inshore craft were based. Although the river was an extremely busy waterway, with an array of different sailing craft working there – the best known probably being the gaff-rigged *fragata* – there were only a few types of sea fishing craft. Smaller craft worked much further up-river, as they did in all the rivers, and the small beach boats working off the beach at Caparica were called *chata*. We've already seen the *saveiro meia-lua*, the surf-launched beach-boats that were unique to the Costa da Caparica in this part of the country and which set drag-nets close inshore. Sardine boats were probably in the majority and Cascais

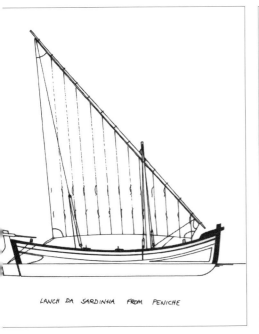

LANCH DA SARDINHA FROM PENICHE

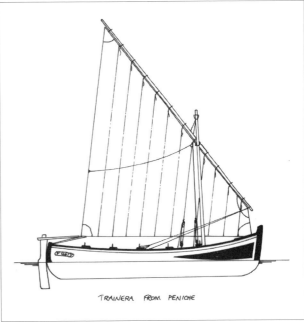

TRAINERA FROM PENICHE

Sailing *traineras* at Peniche similar to the Vigo types.

had its own fleet of lateen-rigged *canoa cacadeiras* that fished up and down the coast and off the Algarve. Although usually setting two lateen sails, they sometimes had a lug mizzen, or even a small sprit jigger. The spritsail was common to Sesimbra, a small yet busy fishing village in a sheltered bay close to Cabo Espichel. The biggest of Sesimbra's boats was the *barca da pesca do alto*, literally a 'high seas fishing boat'. These two-masted, sprit-rigged vessels were shallow drafted, yet sailed as far south as Cabo de Sao Vicente at Portugal's south-west tip. They had upright stems and sternposts, long flush decks and were crewed with up to eight or ten men. When fishing, the masts were lowered and the six oars used to haul around a seine-net. Other spritsail boats worked close inshore, either with *armacao* nets or long-lines. The *canoa da picada* was a sardine boat of Sesimbra and Cascais that had a huge lateen sail, and whose name came from the sardines that were cured in salt aboard. Another Sesimbra boat, simply called a *barco de Sesimbra*, was an open boat with oars and two elegant lateen sails.

The River Sado estuary has been an important fish and salt area for at least seven hundred years. Various types of boat emerged for both fishing and transporting the salt. The three most notable are the *hiate*, the *laitau* and the *galeao*, the latter of which – the *galeao de pesca* – was a lateen-rigged vessel which first transported fish until, in 1890, seining for sardines was introduced from Isla Christina, just over the Spanish border to the south. These speedy little craft, with a crew of up to thirty-four fishermen, supplied the boom in canning the area experienced after a lull in the fortunes of the Breton sardine canners at the turn of the twentieth century. A gaff-rigged version appeared after about 1925 and carried salt until the advent of lorries made this unprofitable. Several examples of these *galeaos* remain, having been restored and sailed once more, one of a very few traditional Portuguese boat types to have survived afloat.

One of the most spectacular of all Portuguese fishing craft were the *muletas* from the River Tejo, based specifically from the ports of Cascais, Seixal and Barreiro. These are not to be confused with the *muletas* or *bateiras*, the Mesopotamian designs with a single lateen sail of Figueira de Foz and Buarco in the northern part of the country. Said to resemble a Norman ship of the thirteenth century, the *muletas* of the River Tejo area were flat-bottomed with extremely rounded stems and sternposts. The stem also had what can only look like shark's teeth protruding forwards. They were open boats that were used to drag a special trawl-net called a *tartaranha*, and they had a glorious assortment of sails to produce the power needed for this heavy net. The mainsail was a big lateen, but both forward and at the stern there were spars called *batelos* (basically a bowsprit and bumpkin) that secured various differently shaped sails, from two triangular sails aft to two absolutely square sails hung off the extreme edge of the bowsprit. The resultant rig is what gave it its distinctiveness, making it the most elaborate and famous of Portuguese traditional craft.

The Algarve region is probably the best known part of the country to the British. Its lovely sandy beaches act as a magnet to the winter-weary who go there. However, as hardly anything remains of its once vibrant beach fishery, it is doubtful whether more than a handful of folk have seen any of the several regional craft that were once common. The *caique* is the first craft that comes to mind. These two-masted

SARDINE BOAT FROM CASCAIS

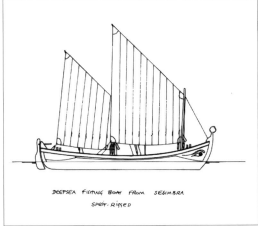

DEEPSEA FISHING BOAT FROM SESIMBRA
SPRIT-RIGGED

GALEÃO FROM SESIMBRA

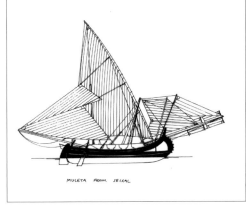

MULETA FROM SEIXAL

lateen-rigged craft were typical of other boats of the Eastern Mediterranean, especially around the River Nile. However, only the smaller *caiques* fished, working out of Olhão and Tavira. Olhao is the largest fishing port in the Algarve and is also home to the 'amazing poodle of the Algarve', a muscular and strong dog, which was also a valuable assistant to the fishermen, for whom it dives into the sea to a depth of over four metres to guide the fish into the nets. These aquatic poodles were abandoned in the 1950s.

The main fishing craft of the southern coast are all canoe-like, with flat-bottomed, double-ended hulls. The *enviada* were lateen-rigged boats that were able to work off the beaches. The word '*enviada*' means literally 'conveyor' as these craft were often used to unload cargo-carrying boats. They were used to set beach seines – *xavega* – or for fishing for tuna. The transom-sterned *enviada do atum*, as it was called, had a keel and, unlike most Portuguese traditional craft, had a small aft cabin for accommodation. As well as the mainsail, it set a foresail on a bowsprit and a mizzen spritsail sheeted to a bumpkin. Again, these boats were used to transport fish from port to port, or from boat to port. Others, smaller, open, flat-bottomed, bluff in the bow and with a raised, rounded stern, worked *xavega* nets with up to six rowing points. The high beak at the bow is reminiscent of many other Mediterranean boats. Today's motorised craft, like

Typical beach boats of the Algarve in the 1960s. Although most were engined, the rig was retained on some craft.

The same beach in 1999. Outboard motors have replaced rigs and the fleet had reduced by probably 80 per cent.

those seen by many visitors on the beach at Albufeira, hardly resemble the older boats of before, with their lean hulls and potent lateen rigs. But then again, that's a story we've seen almost everywhere on our journey around Europe's fish-rich coasts.

And what of fishing in Portugal in more recent times? Some 10,000 boats were still fishing up to a few years ago and many of these are of a traditional design, although many of these had previously been decommissioned and scrapped, and more since. More common are the

ENVIADA

trawlers with high flaring bows that are ubiquitous all along the western seaboard of Europe. On the south coast transom-sterned trawlers of up to about 12 metres work the near offshore. The main deep-sea fishing harbours are Leixoes and Sines, although the western harbours have retained active inshore fleets. Imports of fish from Spain, though, are rumoured to be contributing to the winding down of catches. But one thing is for certain in Portugal, that while steel vessels are being built, some of which are for British customers, wooden boatbuilding is alive. For example, Sagres, where Henry the Navigator built his caravels 'to see what lay beyond the Canaries and Cape Bojador', still has a boatyard building wooden trawlers. Traditional boatbuilding skills are alive around Aveiro, while Vila do Conde is still active with its seine-boats. Indeed, most of the fishing communities have facilities for, if not the building of, then the repairing of wooden craft, which just goes to show that wooden boatbuilding is a skill that isn't going to quickly go away.

SOURCES

The Maritime Museum at Lisbon has a fantastic collection of models of fishing craft. The refurbished Maritime Museum at Ilhavo has a fine collection of boats and information on local fishing folk and their craft. A dedicated fishing museum opened at Praia de Agudo, south of Oporto, in 1999. It houses an aquarium and exhibits the equipment and traditional arts of the local small-scale fishery. Two National Maritime Monographs cover aspects of Portuguese fishing craft – *Boats of the Lisbon River* by Manuel Leitao (No. 34) and *The Decline of Portuguese Regional Boats* by Octavio Lixa Filgueras (No. 47). Both have extensive bibliographies, although most are written in Portuguese. Several recent books in Portuguese also illustrate the variety of fishing craft. *Fishing Boats of the World*, edited by Fishing News, 1996, has a couple of articles in English.

CHAPTER 14

The Atlantic Islands

The Canaries, Madeira and Azores

Portugal also has two autonomous regions, both of which have a history of fishing. These are the islands of the Azores and Madeira, which lie out into the Atlantic Ocean. Not surprisingly, given their positions, both have a long history of both whaling and fishing. In the Azores hundreds of open gaff-rigged whale-boats – *baleeira* in Portuguese – chased the mammals when they came close inshore. Once the boat had arrived at a chosen ground, the mainsail and the jib were furled and the six oars used to work the boat. Up to ten crew, including three harpooners, worked the vessel.

The Azores were also home to a thriving sprat fishery, in which similarly shaped boats participated. However, they had a distinctively different rig in that they set a forward lateen sail on a mast placed well forward and a Bermudan mizzen sail on another mast set about two thirds back from the bow. Again, they were rowed once they'd arrived at the fishing ground, with upwards of ten crew and six oars.

Whales were also much sought after in Madeira. In the small town of Camara de Lobos, on the south coast, a typical whaler was a lug-rigged decked vessel about twenty-five feet long, with a square-headed sail. Although whaling persevered longest in both the Azores and Madeira than anyway else on the Iberian Peninsula, their use largely disappeared in the twentieth century.

The Canary Islands consist of seven islands in the main: Lanzarote, Fuerteventura, Gran Canaria, Tenerife, La Palma, Gomera and Hierra, all of which had a small-scale fishing industry that supported the inhabitants' needs throughout the year. It's not an exaggeration to say that the advent of tourism largely destroyed this artisan fishery. However, some remnants of that fishery do still remain.

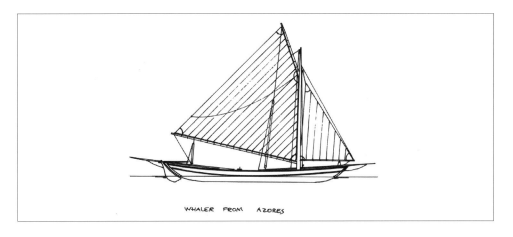

Whale boats – *baleeira* – from the Azores. These were long and narrow and very fast.

Sprat boats were a bit smaller but equally fast.

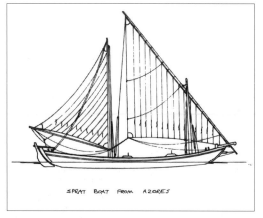

SPRAT BOAT FROM AZORES

WHALER FROM MADEIRA

Madeira in 1876. The beach is a hive of activity. In the foreground at least four tuna boats can be seen while the larger boats seem to be in the process of being scrapped. Judging by the ship's boiler, steamers have also been scrapped here. The two huge wooden capstans are equally intriguing.

The main catch was *bocinegro*, *cabrilla*, *sama*, *sargo* and *mero* (sea bass). But the most prized catch of all was the *vieja*, the parrot fish that is traditionally caught using a bamboo *cana*, a trap, with a special crab as bait. Nets are seldom used. Only when the sardines are in early summer do they set them. Except, that is, at fiesta time in August when all the fishers of the village sail out two days before with nets to catch 7,000 kg of fish to cook and give to those clusters of locals and tourists who gather around the harbours.

Late at night the fishermen still tell their stories and sing songs. One is '*Este noche non allumbra, la farola del mar*', which tells of the lighthouse running out of gas so there's no light tonight to lead the fishermen home. I learnt it when spending six months in Lanzarote nearly thirty years ago and have remembered it ever since. It was said at the time that it related to the lighthouse close by to Playa Blanca, on the south of Lanzarote.

El Cotillo is a small village on the west side of Fuerteventura and is typical of what existed before tourism brought its hotels, foreign-owned villas and money. Today, it has a new harbour which was built about 1990 and now offers the fishermen a deepwater berth for their small, open traditional craft. Thirty years ago it was very

Clearing the nets at El Cotillo in the 1970s. The boats have been motorised and the hull form is fuller but otherwise little else has changed.

The fleet at El Cotillo drawn up in the 1960s, before tourism arrived.

As elsewhere, eventually the bigger motor boats arrived in the Canaries from the mainland, this one being from Santander. The islanders complained that their livelihoods were being taken away but, as usual, no one listened.

different, although the fishermen still survived through fishing. Ten years ago there were some sixteen boats still working from there, with another twenty in Corralejo and fourteen in Puerto del Rosario, the island's capital. All the fishermen were members of the Cofradia de Pescadores – Fishermen's Association – which seemed to be the only voice fishermen had left. Their boats were small open vessels, pointed at both ends and of two distinct shapes. Firstly there were the narrow *barquillos*, the original boats that were once rigged with a lateen sail for getting out to the rich fishing grounds off the island. These were kept on the beach at La Caleta, otherwise known as the old harbour, brought up high in winter, away from the Atlantic rollers that can move huge boulders from the beach into their houses. Fuerteventura has some of the best fishing in all the islands, a fact due to the relatively shallow water offshore. Although these boats aren't rigged today, their shape remains largely unchanged.

The more modern boats are called *faluas*, and these are fuller, although otherwise much the same, due to the addition of engines. Up to about 1940 engines were hardly known, sail being the only option when you're living at the edge of the world.

There was still a boatbuilder in El Cotillo a few years ago. Ildefonso Hierro worked by supplying the tuna fishermen of the south with vessels. His father Santiago built some 500 fishing boats in 30 years. Now the need has disappeared.

El Cotillo wasn't always about fish and tourism. It grew out of the trade in lime, necessary for water purification, building purposes and lime-wash to paint these houses. Four calcinating ovens sit above the new harbour as mementoes. It seems that there are still 149 relics of these ovens on the island. Lime was shipped out from the old quay, the remains of which can be seen by the beaches with the stone-shelters. Then, between the wars, the population moved away, leaving only two or three families behind. After the Second World War they returned, to eke out a living from fish. Their small lateen-rigged boats were beached at the old harbour, brought up high in winter. When the *falua* arrived, fishing became more of a business as the fuel had to be paid for. For a while tourism created a market for fish and times were good. Then the bigger European trawlers arrived, denuding their seas of fish and, with the winter sun attracting thousands from the north of Europe, fishing was suddenly not the best option.

SOURCES

There's a Museum of Whaling in Madeira, from which some information came. Various books give accounts of whaling, especially from the Azores. In the Canaries most of the information came from personal communication with the boatbuilder Ildefonso Hierro and the 70-year old fisherman Chano.

PART THREE
THE MEDITERRANEAN

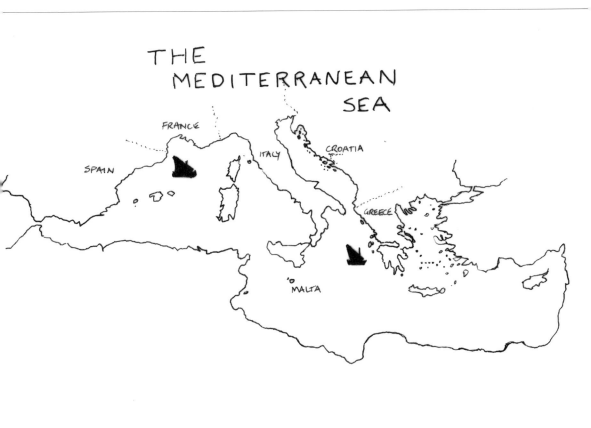

MEDITERRANEAN COASTS
OF SPAIN + FRANCE

FRANCE

SÈTE

MARSEILLE
CASSIS
NICE

Golfe du Lion

ITALY

Ligurian Sea

SPAIN

BARCELONA

CORSICA

VALENCIA

Islas Baleares

MALAGA

GIBRALTAR

CHAPTER 15

The Mediterranean Coasts of Spain and France

Gibraltar to the Gulf of Genoa

SPAIN

The Mediterranean traditions are very different to those from northern Europe. Warm, ideal climates throughout the majority of the year produce a very different array of craft, the majority of which are open boats working off beaches or sheltered bays. However, more profound are the influences from ancient times, which resulted in very different, and innovative at the time, characteristics that define Mediterranean fishing craft. Today huge numbers of replica boats, restorations or simply the later generation of motorised craft crowd the busy Mediterranean harbours, vying with the modern plastic yachts for the visitors' attention. In some of the villages, with dramatic backdrops of tall, stone-coloured buildings clinging to cliff-faces, the older craft evoke memories of days long gone when the sea was the prime way of communicating with the outside world as well as producing a healthy sustenance.

The coasts of southern Spain and France are less rugged and dramatic than their Atlantic counterparts. Gentle sandy beaches – so popular with the hordes of summer visitors – have been perfect for beaching craft. Harbours are numerous, protected by low headlands, and fish were once prolific.

The Mediterranean is home to the lateen rig, with its *tartanes*, *feluccas* and *barquettes*. The term '*barquette*' seems to be a loose term for Spanish and French lateen-rigged craft. They are double-ended craft, between about 6 and 7 metres long, with flat floors to enable them to sit upright, although they also have shallow bilge keels either side. Like most Mediterranean craft, they have the phallic beak – the stem sticking up above deck level which, so it's said, was banged when coming onto the beach to attract the attention of the fish buyers – and a mast that characteristically leans forward. In bygone days they operated seines, although latterly they have been used for all sorts of fishing, from sardine netting to squid fishing with bright lights hung out over the stern to attract the fish.

Catalan Bay was traditionally home to the beach-based fisherman of Gibraltar and numerous open boats still sit on the beach. Just over the border, there's a fleet of similar craft at La Linea, where the fishermen still use hand-operated windlasses to haul their boats up and down the beach. Watching a couple of fishermen walk round and round, pushing the heavy beam, while another guided his boat up the pebbles, I was reminded just how some parts of the Mediterranean have hardly altered, while other bits are hideous masses of concrete blocks and marinas. However, it must be noted that the vast majority of these craft have engines at the expense of the rig. Who can blame the fishermen for adapting to twentieth-century advantages?

However, before we fully consider the fishing craft of the Mediterranean coast, we must first mention the early caravels, which were oared craft of a length up to about 9 metres that fished with a seine-net. They were long, narrow craft with up to five rowing points. Although said to be one of the oldest types of fishing craft along this coast, their use survived into the twentieth century. Other variations of the *barquette* are as often as not named after the method of fishing the boat was used for. Thus a *palangrero* is a lateen-rigged boat which used paternoster lines and was decked over except for a central hatchway through which the mast was mounted and access to below was had. A *barcas de jábega*, then, is a seine-net boat from the Malaga region (*jábega* is the net) which dates back at least 200 years. They were probably even older, as it has been suggested that they are one of the oldest boat types in the Mediterranean, and that they were introduced into its western parts by the Phoenicians, who also brought with them the oculus that was normally painted on both sides of the boat's bow. Being of 2–3 tons and between 7 and 9 metres in length, when fishing the boats carried seven or nine oarsmen and had three further crew: the skipper, who steered with a long oar, the net shooter and net controller. They seldom worked more than a mile out from the shore. There are still many examples sailing in the waters of Malaga, although not in the numbers that there once were.

These craft were found to be easy to install motors on and so their shape hardly altered, but for the normal fullness and alterations to the sternpost configuration to accept a shaft and propeller. They became fully decked and were again characterised by the method of fishing so that a *bote de luz o 'gussi'* was a squid boat with one or more bright lines hung over the side.

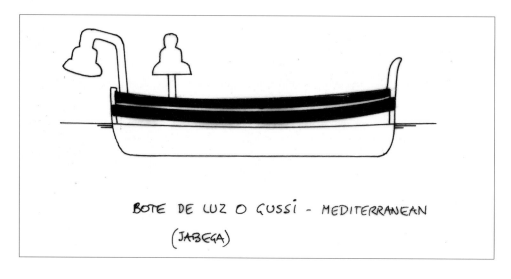

BOTE DE LUZ O GUSSI - MEDITERRANEAN
(JABEGA)

Catalane Bay, Gibraltar, with its traditional lateen-rigged craft.

The *barca de jábega* is the seine-net boat of this coast, seen here at Malaga.

Several miles south of the orange-exporting city of Valencia, on Spain's Mediterranean coast, is the freshwater lake L'Albufera, which is renowned for its terrific sunset views and bird life. It was also home once to the *llagot*, a double-ended, flat-bottomed canoe that is now exclusively worked under oar. A century ago, though, the lake was much larger but has shrunk due to land reclamation, thus reducing the fisheries, although it is also said that pollution over the last fifty years has also contributed to the loss of fish. Eels were once extremely rich in supply and larger *llagots* worked under a lateen rig but the sailing craft are no longer considered necessary. Today there are said to be some 400 fishermen still catching eels, which are cooked locally as *all i pebre de anguilas* – a spicy eel dish. The tradition of having their fishing grounds (*calaes*) assigning on the second Sunday in July by drawing lots still survives, as does the *Festa del Cristo de la Salud* on 4 August, when a procession of boats carries the crucifix from the church of El Palmar across the lake and back, the boats being covered in flowers. Today many of these boats are motorised and partially decked and used for taking visitors around the lake to see the birds or the sunset. However, while passing through a few years ago, I did find several examples of the rowed version of the *llagots*.

Several other sea fishing types are worthy of a mention. In the Balearic Islands, the *llaud* was a typical fishing craft with one lateen sail, peaked high, that seems to have originated from the Middle Ages. These appear to have been smaller versions of the three-masted *barca mallorquina*, large trading vessels that worked among the islands. Because of the proximity to Catalunya, the *llauds* also work on the mainland. The vertical stemmed *Barca de Bou* were trawlers that worked in pairs which resembled pairs of oxen ('Bou' means beef). These were lateen-rigged upon a mast that raked sharply forward, a characteristic said to aid going about. After about 1920, motors were installed which allowed single boats to pull their own trawls using otter boards. One surviving example, the 1924-built '*Balear*', is owned, and has been restored, by the state of Mallorca.

Sardinals were sardine catching craft while *quillats* were small beach boats working from Barcelona and the beaches of the vicinity. Both were lateen-rigged and both survived the transition to motorisation.

But it is the lateen-rigged *barca catalan* that today rules supreme along the coast from Barcelona northwards and, indeed, well into the French part of the region. Whereas in fact they don't differ much from the *barquettes* seen all along the coast, there does seem to be a semblance in restoring these craft on this stretch of coast. This is the Costa Brava, the wild coast, which today remains less spoilt than the coast in both directions. Rarely does a beach not have an example of these fine craft, and in Spain they are mostly painted white. Over the border with France they become the *barque catalane* and are gaily painted in all colours. Regardless of the border, they were a Catalonian boat.

FRANCE

Between the anchovy port of Collioure, near to the Spanish–French border, and the Italian Ligurian coast, some fourteen individual fishing boat types have been identified,

LLAGUT FROM LAKE ALBUFERA

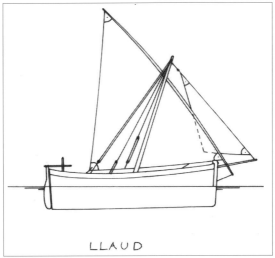

LLAUD

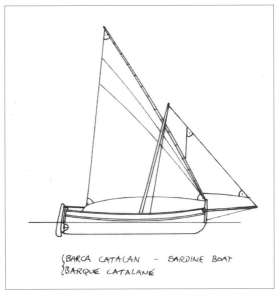

(BARCA CATALAN - SARDINE BOAT
{BARQUE CATALANE

quite a large number for a relatively short coastline of some 350 miles. The diversity is equally interesting.

To the west, as mentioned, it is the *barque catalane*, the typical boat of small harbours such as Collioure, Banyuis and Roussillon. These were primarily anchovy boats, drift-netting for the small fish between spring and autumn, though they also fished the sardines and occasional trawled in pairs in winter. They are generally regarded as the prettiest littoral fishing boat of the Mediterranean because of their sleek shape and fine line in the water. They had fine entries and rounded, spacious hulls and were double-enders with *capians*, the extended stem head beaks, which were often moulded in a very erotically explicit manner, phalluses, as a sign of offerings to Neptune (and his fertility) ensuring safe sailing and good fishing, as well as alerting the fish buyers. It is said that these were introduced by the Greeks and are seen in many parts of the Mediterranean. The boats ranged from 8–12 metres and set a huge single lateen sail – *la mestre* – bent onto a yard that consisted of two spars – the *penn* and *car* – bound together which could be over 15 metres in length. Again, the midship-positioned mast raked forward. Sailing windward the fore end of this was sheeted tight against the *capian*, though when downwind this was freed off. The sail was able to traverse right round the mast. It is said that they were easy to sail though, from pictures, they do not appear that simple! Crewed by three to five men, they ventured far and wide, as far as the Ligurian Gulf and Sardinia. Many have been restored and sail during the summer.

Along the fringes of the Golfe du Lion, between the Spanish border and Marseille, are inland lagoons or lakes called *les etangs*. Thus, each had fishing craft that worked in the same way as the *llagots* in lake L'Albufera. These slab-sided small craft, some flat-bottomed, are variously named *bétou, bétoune, nacelle, barquet, négafol, nagachin* and *bette*, depending on which *etang* they worked upon, such as those at Leucate, Bages, Gruissan, Thau, Vic, Vaccarés and Berre. Most were lateen-rigged, though the sprit-rigged *bétou de Gruissan* was an exception. Some, such as the *bettes*, were built by boatbuilders to order by the *pan*, a unit of measure equal to 25 cms.

The *bateau boeuf de Sete et Martigues* was named in the same way as the *Barca de Bou* in Mallorca in that these boats worked pair trawls resembling oxen. The boat is said to have evolved from the *tartans* of the nineteenth century. It was a heavily built, solid vessel and consequently slow.

The *barquette marseillaise* is probably a version of the most

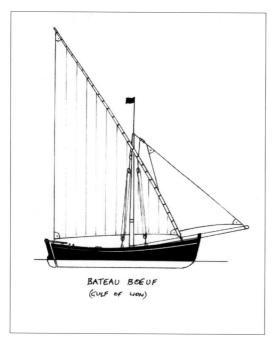

BATEAU BOEUF
(GULF OF LION)

widespread of all boats in the Mediterranean. It appeared in its present form in the second half of the nineteenth century, when Italian immigrants moved from around the coast of the Italian region of Campania, including the coast of the Gulf of Gaete, from the Naples area and the Amalfi coast. These people were mostly fishermen and boatbuilders, so they soon began building craft modelled on the *gozzi* they fished with (of which we will learn more in the next chapter). The *barquette marseillaise*, built by boatbuilders with names such as Ruoppolo, De Stephano, Autiero, Aversa, Trapani and Battofero, were of a light construction, beamy though very fine at the entry and exit and rigged with one lateen sail. They were almost symmetrical and often decked with a large hatch running amidships. They had a low freeboard for fishing and, with their good lines, were regarded as a very pretty vessel. They also had a *capian*. The boats themselves were also regarded as being very different to the other boats of the area. The *barquette marseillaise gangui* was a larger version, built to trawl for flatfish, and seemed to originate from Estaque, one particular affluent area of Marseilles.

The *Mourre de pouar*, sometimes referred to as the *Mourre de porc*, was a very beamy boat with a pointed stern, generally known as a *Pointu*, as are all the craft along the French Mediterranean if they are double-ended. Indeed, there is much discussion about whether a boat is a *Pointu* or a 'barque' (see www.carenes-association.org). To confuse matters, the *Barquette marseillaise* is generally regarded as being a *Pointu*! One famous *pointu* is '*Piou-Piou*', a vessel built in 1910 by Feraud of St Tropez and which was owned by Bridgitte Bardot for several years and is still afloat.

The *Mourre du pouar* were common around the mouth of the River Rhone, from Grau-du-Roi down as far as Toulon and La Ciotat. Their name translates to 'pig's snout' or 'pig-faced' and comes from the fact that, when viewed from ahead of the bow,

BARQUETTE MARSEILLAISE

Above right: A *barquette marseillaise* under sail.

the fullness of the hull and the spur extending above the bowsprit gives the impression of looking closely at the snout of an old boar. However, they were fast boats and in 1818 were adopted as pilot boats because of this and their manoeuvrability, which meant the pilot's boy was capable of sailing back to base alone once the pilot had been dropped off aboard an incoming vessel.

La Ciotat is also home to the *Ciotaden*, a version of the *barquette de La Ciotat*, which is itself a form of the Italian *gozzi*. However, over the years this has been adapted for racing. The waterline length was increased without increasing the length at deck, thus producing a vessel with both the stem and sternpost raking outwards. However, there is no evidence that these vessels fished commercially.

The *bateau a livarde du Lavandou et de Saint-Tropez* is unusual in that it is sprit-rigged, unlike the majority of Mediterranean craft with their lateen-rigs. This rig is by its very nature much easier to raise and lower and the sail can be quickly furled. The boats were light, between 5 and 7 metres, and thus, with the rig, were manoeuvrable so that they could fish close to the rocks. They generally set gill-nets, trammel-nets, long-lines and fish traps.

The *gourse de Nice et Cannes* is a beach boat, often only rowed, which worked the eastern extreme of the French coast. They were up to about 6 metres in length and before tourism flooded the Riviera coast with buildings, marinas and private beaches, were commonplace anywhere between Cannes and the Italian border at Menton. Nowadays, those that do remain, like almost everywhere along the European Mediterranean coast, are fuller in hull form because of motorisation. However, with regard to the gourse, these are, as has already been mentioned, in fact Italian craft which will be described in the next chapter.

Today's craft fishing this coast are like those elsewhere around the Mediterranean, with wheelhouses, powerful engines and winches and transom sterns giving ample working room aft. However, the vast majority are still wooden, built at various towns along the coast and up to 25 metres in length. Fishing, traditionally the fare of the local beachmen, has quickly passing into the hands of the few with their gigantic trawlers, capable of sweeping up everything in sight. Again this is the typical situation, and is a reflection of the ill-conceived Common Fisheries Policy which does just about everything except ensure sustainability! However you look at fishing, political decisions cannot change the one underlying factor that besets the industry. Too many fishing boats are chasing a decreasing fish stock. Dispose of the largest boats that catch the highest proportion of fish, then the coastal fishers can be left to pick up the pieces in their small coastal craft – just like they've been doing for centuries.

SOURCES

For Spain much of the information came from the Maritime Museum at Barcelona, although there is a new *Museu de la Pesca* at Palamos on the Coast Brava. For France, Chasse Marée's *Guide des Voiliers de Peche* has been vital.

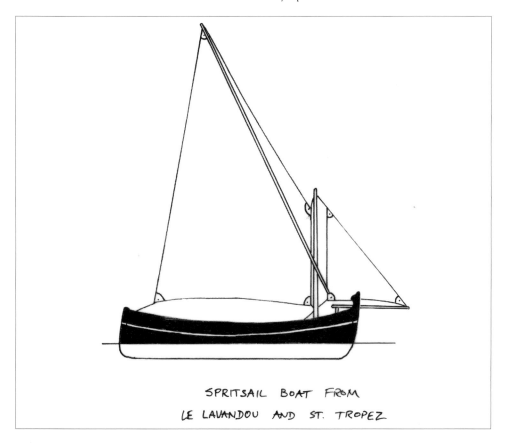

SPRITSAIL BOAT FROM
LE LAVANDOU AND ST. TROPEZ

Bateaux a livarde at Saint Tropez.

CHAPTER 16

Italy

The Gulf of Genoa to the Gulf of Venice

MARE LIGURE E MARE TIRRENO

When one crosses the border from France into Italy, the discerning ethnologist will notice little change in the design of fishing craft in the immediate vicinity. The double-ended *barquettes* of Marseille, the *Pointu* and other the boats of the St Tropez peninsula and, especially, the *gourse* of Nice all bear certain resemblances to the *gozzo* of the Ligurian coast. It's the same blur of influence, as we saw in the previous chapter, where the change from *barca catalan* to *barque catalane* is as obscure at the Spanish/French border. Fishermen have never been renowned for their respect of these physical borders that the sea has for all times ignored.

The *gozzo* is said to be one of the most common of boat types in the Mediterranean. As well in southern France, its contemporaries can be found further south at Naples and Sicily, in the Balearic Islands (as the *llaud*), in Catalonia (as the *barca*), Malta (the *Gozo* boat), Greece (the *caique*), in Dalmatia (the *gaeta*) and the north Adriatic (the *guzzo*). On the Italian Ligurian coast, which can be said to extend from the French border to about the island of Elba, there were different types and shapes of *gozzi liguri*, which in turn depended on their use, where they were built and the beach or harbour they worked from. As an example, their

sizes ranged from about 5 metres up
to 10 metres. Today they are still
actively sailed or rowed, depending
on size, and various regattas are
keenly competed.

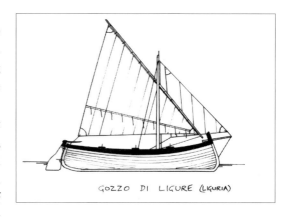

It has been said that this coast
is not renowned for its fish, meat,
dairy or grain production, its food
consisting of mainly pasta and
stuffed vegetables. It also has one of
the highest rates of life expectancy
in Italy, according to one writer. Another, Claudia Roden, suggests that the area's
love of vegetables is primarily due to the hankerings of the sailors when home from
the sea, tired of a monotonous diet of fish and salt beef. Genoa was, after all, one of
four maritime centres in what is now modern Italy – the others being Pisa, Venice and
Amalfi. Furthermore, Camogli became known as 'the port of a thousand white sails',
reflecting its obvious maritime importance. However, in direct contrast to the above
assertions, others report that the villages of the Cinque Terre grew out of places where
fishermen have fished for over a thousand years while, at the same time, being the
haunts of pirates up to the eighteenth century. *Gozzi* were also used to transport their
crops, primarily olives, to market.

What is sure is that the *gozzi* were either sprit or lateen-rigged. Some were rounder
than others, depending on the degree of shelter at home. On the western edge of
Liguria the boats often were built with a backward-raking stem (*alla cornigiotta*)
which is said to originate from Cornigliano Ligure, slightly west of Genoa, where the

Gozzi type boats of the Gulf of Liguria pulled up in front of the abbey of San Fruttuoso, near
Camogli. The nets are drying on lines suspended from the hillside and the caves beneath the
abbey used by the fishermen.

Above: Camogli harbour, with a mixture of trading and fishing craft.

Left: Camogli in 2005. The *gozzo* has not changed throughout generations.

boats had to be launched into surf. The hollow shape of the stem ensured that no water was shipped aboard. Other *gozzi* were suited to be moved about on greased wooden posts laid on the beach. Since motorisation in the 1920s, the traditional *gozzo* had a forward-raking stem (*alla catalana*) instead of a vertical one. The name itself surely reflects the similarities between these craft and those from further west. In nearly all cases, they are said to have been built with acacia frames and pine planking.

The last true *gozzo* is said to have been the '*Padre Santo*', built in 1938 and, until recently anyway, still surviving at Santa Margherita Ligure. A recent visit failed to unearth it but there is no doubt that the shape lives on in more modern craft, built from modern materials and, mostly, fitted with outboard brackets in place of a rig.

Another vessel, said to be older than the *gozzi* and to be better suited to fishing, is the *gondola* of the Ligurian and Tuscan coasts. They were originally sprit-rigged and later adopted the lateen. In 1746 there were twenty-three of these vessels in existence. They were without much sheer, rarely exceeding 8 metres, and yet could easily carry seven crew. They were mostly used for inshore fishing. The *rivanetto* is another lateen-rigged vessel, of about 9 metres, used for working off the beaches of the Tuscany coast, albeit finer in body section than the *gozzi*. These are a form of the *latini*, which seems to be a very loose term for all lateen-rigged vessels that work off these north-western coasts, especially for inshore fishing. The *leudo* are far larger vessels which appear primarily to have been used for carrying cargo. These evolved out of the older *liuto* that date back to the seventeenth century.

As already mentioned, the *gozzo* was also to be found around the Gulf of Naples, where it was known as the *gozzo di Sorrento*. This was a much sharper and narrower boat than its counterpart to the north, very flat and without sheer, although the body section was full and almost flat-bottomed. They were sprit-rigged with a jib on a long bowsprit. They were seldom longer than 8 metres. One example was discovered recently on the beach at Cetera, on the Amalfi coast.

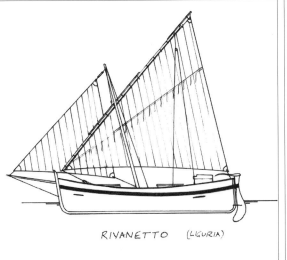

RIVANETTO   (LIGURIA)

GOZZO DI SORRENTO

The *gozzo di Trapani* was another such craft whose similarity is only in name. Worked exclusively on the west coast of Sicily, these beamy little craft fished outside the tuna season. For the tuna, one of the most important fisheries in the Mediterranean, the Sicilian fishermen of the west worked the *barche di tonnara*, literally the 'trap-net boats'. The tuna trap is an extremely large and complicated structure weighing several hundred tons. The season lasts from May until June, at which time the fish spawn in the warm waters of Sicily. The use of these trap-nets is said to date back to their introduction by the Arabs in the ninth century. *La mattanza* is the ritual catch and the tradition is especially deep-rooted in the culture of the Isola Egadi.

Two types of boat are used, which are then beached for the remainder of the year outside the season. Large boats, with two masts with lifting derricks, set the net in about five hours. These are transom-sterned, flat-bottomed boats that are large enough to carry the heavy net and later to transport the tuna back to shore. Smaller boats, double-enders with no rake in the sternpost and rigged with lateen sails, were used to attend the openings in the net and to literally chase the fish into the inner chamber of the net known as the *camera del morte*. However, when chasing, they are only propelled by six oars. Some sixty fishermen participate in the fishery. Once the tuna are captured, on the instructions of the *rais*, the chief fisherman, the fish are killed by stabbing and removed using long-shafted gaff hooks. The slaughter of the fish and the sea turning red is not for the faint-hearted! There's a saying here that 'tuna fishing shortens your arms and silences your tongue'. Both types of boat are painted black throughout.

Sicily is also renowned for its sardines, as is much of the west coast of Italy, and the *sardara* was a version of the typical Sicilian boat, double-ended with curved ends and some 6 metres in length with flat floors amidships and a fine entry and exit. The mast was set almost amidships and lowered during fishing, at which time the three rowing points were used. Rigged in the main with a lateen sail, these boats were low in the water. Another vessel was the *varca ustanisa*, another lateen-rigged boat from the south-east of the island, a sort of all-round boat. Both these boats, in fact, are fairly typical of the Italian west coast boats and can, tenuously, be said to lie somewhere between the traditional west coast *gozzo* and the flatter and bluffer Adriatic craft. The sprit-rigged *varca di conzu* was simply a variation of the *sardara*. Another Sicilian boat is the canoe-sterned *luntro* which was used for swordfish fishing in the Straits of Messina, another fishery that is centuries old. Rowed by four crew members, there was a lookout stationed up the mast while the man on the prow would harpoon the fish. They usually worked in tandem with a searcher boat. After motorisation, a design was developed that combined both operations, allowing access to deeper waters in and around the Isola Lipari to the north, these having mast pylons over 18 metres high and walkways at the front extending out at least 10 metres. However, fleets of fast, 30-ton feluccas that are equipped with harpoon guns have superseded these.

Likewise, the coast of Sardinia had its own boats that were largely variations of mainland types. On the west side, the *spagnoletta di Alghero* resemble the *gozzi ligure*, while the *schifetto di Carloforte* is a deeper vessel unsuited to beaching. Both

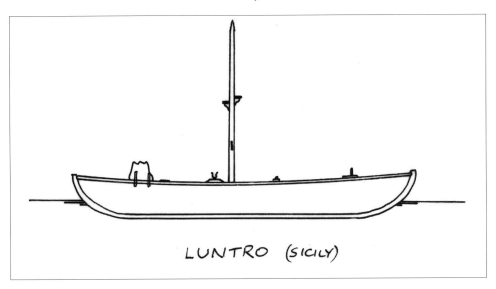

LUNTRO (SICILY)

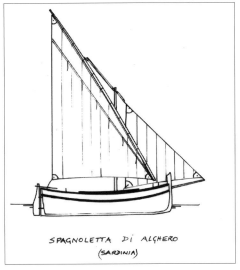

SPAGNOLETTA DI ALGHERO
(SARDINIA)

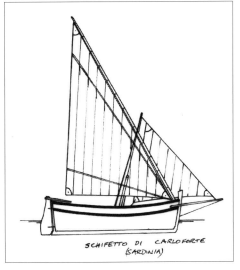

SCHIFETTO DI CARLOFORTE
(SARDINIA)

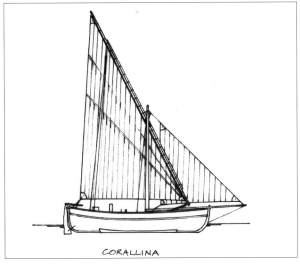

CORALLINA

were lateen-rigged. Fishing is said to have been carried out by people from outside of Sardinia where tuna was once caught off the north coast but no longer survives. The cultivation of mussels and shellfish is nowadays the chief mode of fishing. However, the *fassone*, the reed boat of the waters in and around the Gulf of Oristano, is unique in Italy. The reeds are collected locally and allowed to dry before being bundled and tied to make a one-man boat capable of lasting up to a year. They were used primarily for catching sardines in the shallow water of the lagoons (*stagno*).

Two other vessels in common usage were the *lancia* of Pantelleria, an island close to North Africa, noteworthy for its transom stern, and the *barca di Ponza* from the island of that name in the Tyrrehian Sea, where the inhabitants were one of those groups fishing off Sardinia. And, to complete the picture on the west coast, the *corallina* was specifically a coral-fishing boat, while the *lautello* was used exclusively for sponge fishing.

Today's fishing boats have, as is the case throughout all of Europe, adapted to motorisation so that regional variations have all but disappeared. The shape of the stern has allowed a propeller aperture while the fuller shape has accommodated the weight of the engine and, often, all the fishing gear. Today the *batelo* is ubiquitous in that the term simply refers to the boat. And it is only the discerning eye that can tell one boat apart from another.

MARE IONIO E MARE ADRIATICO

Mazara del Vallo lies on the west coast of Sicily, somewhere near the imaginary border between the Tyrrehian and Ionian Seas, and is one of the country's principal fishing harbours, filled nowadays with large, high-bowed boats that penetrate deep into the Mediterranean Sea in their search for fish. Along the southern coast of Sicily the harbours of Licata and Syracuse are perhaps the main fishing centres. In contrast, the Calabrian coast – the 'toe' of Italy – was renowned for its long association with disease and piracy for centuries after the fall of the Roman Empire, largely due to the coastal plains becoming malarial swamplands. The locals fled inland, away from the coast, so that today no local dish includes fresh fish. In the 1950 Reform Act, land was given to the farm labourers by the landowners, who kept the best ground inland for themselves. Thus, the coastal area was re-inhabited and a fishing industry developed. However, no particular individual type of boat ever developed. Further west, at Taranto, there's a thriving oyster trade in the shallow waters around this ancient port where small, flat-bottomed boats, today with outboards, have worked. Nearby Porto Cesareo and Gallipoli are thriving fishing harbours where the small, beamy boats are similar to those from Sicily, albeit motorised.

The boats of the Adriatic can be said to have been much better researched over the years, in the main by Mario Marzari and Velimir Salamon. To understand in full the variety and development of these craft, it is necessary to jump some 500 miles and consider the northern Adriatic ports of Venice and Trieste. And the one difference immediately noticeable is that, unlike the west coast, many of the east coast boats use the lug rig as against the lateen.

It is firmly believed that the single-masted, lug-rigged *brazzare* originated in Dalmatia in the sixteenth century but they came to frequent every corner of the Adriatic Sea. Used mostly to transport goods, they also fished and gathered sponges like the *lautello* of Sicily. Sizes varied but ranged from about 9 to 14 metres. The *tartana da pesca*, on the other hand, evolving from earlier fourteenth century Venetian craft, reached its maximum potential in the seventeenth and eighteenth centuries, a hundred years before the *brazzare* reached its. These boats had bluff, high stems and similar sterns with little sheer amidships. Their chief port was the town of Chioggia, south of Venice, which today remains one of Italy's main fishing harbours. Although its quay is now lined with upwards of a hundred boats, very few resemble anything like the older vessels. The *tartana da pesca*, similar to the generic *tartana* that worked throughout the Mediterranean, is said to have been the only deep-sea Chioggian fishing vessel in that it worked far offshore. However, it was still flat-bottomed for working in the shallow waters of the lagoons, and had a huge rudder extending below the bottom of the boat to serve as a keel. This could be raised in shallow water. The boat was rigged with two lugsails and, like many other north Adriatic boats, was renowned for its colourful sails. On board, smaller *topo* boats acted as tenders. Various methods of fishing were employed, such as long-lining (*parangal*) or trawling with a *cocchia* net when they worked in pairs. The authorities of Venice at one time declared that these boats were not to fish in the eastern Adriatic but repealed the legislation ten years later when it discovered that the Chioggian fleet had diminished by some thirty-four boats because of the decree.

The *bragozzo* was another distinct Chioggian vessel and was not unlike the *tartana*. It gained precedent after this 1770 decree due to its smaller size and greater suitability for fishing the shallow waters which stretch way off almost the entire east coast of Italy. Again, it was flat-bottomed and recognisable through its massive rudder. It has been said that they were built entirely of oak. They had two lugsails, with the forward one being much smaller and set on a forward-raking mast. Again, their sails were heavily decorated. Surely the shape of the barges surviving today on the canals of Venice must originate in the hulls of these older types.

There are two other, larger types of fishing boat worthy of a quick mention. The *trabaccolo*, a vessel with a high and bluff bow, is normally decorated with 'eyes' in bold relief on the stem, in much the same way as the Greeks and Maltese decorate their boats. Again, they are lug-rigged with two masts and have large rudders. The *bragagna* have been described as the 'most elongated fishing boat of the Adriatic' and were used around the lagoons of Venice. They had three masts with lugsails and were in use from about the sixteenth century up to the nineteenth. They trawled with nets, pushed sideways by the wind.

The already mentioned *topo*, literally translating to 'mouse', became a widely used fishing boat throughout the Venice region. It was half-decked, flat-bottomed and single-masted with a lugsail. Fishing with lines was its normal mode of work. It is said that, together with the *bragozzo*, the *topo* represents the most evolved specimens of flat-bottomed boats.

The *lancia Romagnola* bears no resemblance to the west coast *lancia*. It has been in the past referred to as the Ancona fishing boat, and its true origins come from around that region and its usage never penetrated further north than Romagna. It is said that it represents one of the most recent types of traditional Italian fishing craft yet, in terms of hull decoration when compared with craft from Venice, it was one of the most restrained. The typical *lancia* is about 9 metres long, 2.1 metres in the beam, flat-bottomed for easy beaching with some 60 cm of draft, and has a single lugsail and two foresails set on a bowsprit. Two prime examples still exist at Cervia. The 'Assunta' was built in 1925 at Patrignani di Cattolica and has been fitted with a Farriman Diesel 18 hp motor. The other, the 'Maria', was also built at Cattolica, which is between Rimini and Pesaro, in 1949. Both have brightly coloured sails with individual patterns that are unique to each fisherman. Usually they represent holy scenes and are to enable those relatives on shore to identify the returning boats.

Cesanatico lies a bit south of Cervia and is also home to a fleet of these craft. Thus the summer races are keenly competed, with each saying their craft are the finest. Each year they join in with the 'manifestation of the fishery', at which time the boats sail from Cervia with a local priest or bishop aboard who, once the company is assembled, throws a ring into a group of swimming youngsters. If the ring is caught, the fishery will be bountiful and, if not, some don't even bother. In 1986 the Pope came to officiate and the ring was caught. At his visit he sailed out upon a *bragozzo*. This tradition reaches back to 1545 and, on 1 June 2013, the ring will be tossed into the sea for the 568th time.

Cesanatico also has an outdoor collection of other vessels in use on this coast until modern times. As well as those already mentioned, there's a *battana*, which is regarded as the most economical and easiest boat to build. Numbers were largest in this harbour. They were rigged with one lugsail that could be lowered with ease and had the traditional flat-bottom with large rudder.

The word *paranza* translates to 'the boats which proceed in pairs' and reflects the mode of fishing that they do. They are small copies of the larger *trabaccolo da transporta*, the cargo boats that are the largest of the native boats and work up and

PARANZA

down the coast carrying everything from sand, stone, coal and wood to oranges, flour, melons, wine and olives. The smaller *paranza* have one large lugsail and, as their name suggests, dragged nets in pairs before the onset of the otter trawl. They worked all parts of the Marches and Abruzzi coasts. Two other types that exist but were unseen are the *sanpierota*, a flat-bottomed boat not unlike the British flatner, and the *batelo a pisso*, a smaller version of the *bragozzo*.

South of Ancona the coast is bare and with little fishing activity until arriving at

A gaff-rigged *lancia* returning to Cesenatico in the 1950s. The days of the painted sails had long gone by then.

A small boat – shaped similarly to the *lancia* – being winched along, although the reason is not certain. The water is shallow, so the obvious reason is that it is being hauled ashore, but this doesn't explain why a capstan is aboard such a small boat.

A *bragozzo* entering Cattolica, a line being thrown over to help her sail up the narrow entrance.

Examples of the *sandolos* upon the beach at Rodi Garganico. They were still being used for fishing in 2008.

Internal view of a *sandolo* showing its flat bottom
construction which resembles many other European
vessels of a similar nature.

the Promontorio del Gargano, one of the only remaining wooded areas of Italy. At Rodi
Garganico there are still examples of the *sandolo*, a boat that had been introduced into
the local fishery from the nearby Lake Varano. It, too, was flat-bottomed, and simple
to build, light to use and cheap to build. One was measured on a recent visit and the
resulting drawings are shown here. Although, from photographic evidence, it is clear that
larger boats did work from the tiny harbour, these show strong Greek and other Adriatic
influences, and appear very different from the craft to the north. Today's *batelo*, meaning
'boat', is what the motorised boats that sit on the beach are called by the fishermen. They
are prolific all around the peninsula and down as far as Brindisi and beyond. At Rodi
Garganico they are used only for sailing further out while the *sandolo* are capable of
working up to a mile out, but further south there was no alternative. And, as we've seen
throughout Europe, today's political and corporate pressure, especially upon the fisheries,
has assured that individualism is discouraged while blandness and profit avail.

SOURCES

Vele italiane della costa occidentale by Sergio Bellabarba and Edoardo Guerreri covers
the entire west coast and contains wonderful images of each craft mentioned. *Gozzi di
Liguria* by Giovanni Panella gives a thorough understanding of these particular craft.
Both books are in Italian. On the east coast, the museum at Cesenatico has a wonderful
collection of traditional craft, most of which can be seen afloat in the harbour with their
colourful sails up in summer. For the *bragozzo*, the book *Il Bragozzo* by Mario Marzari
is essential reading. This author, sadly now deceased, was a wealth of information on
Adriatic craft and he produced various other books including *Vele in Adriatica*.

CHAPTER 17

Malta

Malta and Gozo

Malta, along with its sister islands of Gozo and Comino, sits centrally in the Mediterranean, some sixty miles from the nearest other land, and for centuries has been a staging point for vessels traversing from one end of the Mediterranean to the other. With a history of conquests from all directions, her maritime heritage abounds with influence from Europe, North Africa and the Middle East.

The earliest settlers were from Sicily, the nearest neighbour to the north. They arrived in about 5000 BC and were hunters and fishermen who made their homes along the coast. The Phoenicians arrived in 800 BC and although their influence was widespread throughout the Mediterranean and further, it is, as yet, unclear whether this influence affected Maltese working boats.

Sicily again influenced various aspects of life on the islands when, along with Malta, she was under Arab rule in the Middle Ages. It is hardly surprising, then, that Sicilian motivation has left its mark upon the Maltese working boats up to the present day – indeed, the name *luzzu* is probably of Sicilian origin.

The earliest documented fishing boat was the *ferilla*, which originated around Grand Harbour as a small rowing boat. This had always been the principal Maltese fishing boat until the *luzzu* appeared in the early part of the twentieth century, From its first

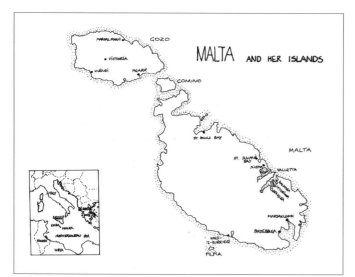

origins as a rowing boat, it was adapted by those who started to go out fishing for profit. These fishermen were said to have been shopkeepers who found it necessary to fish for more than they could eat themselves, the surplus being sold on. This trend towards commercial fishing first arose in the fifteenth century, when Malta was only able to produce a third of its food requirements. The sea was

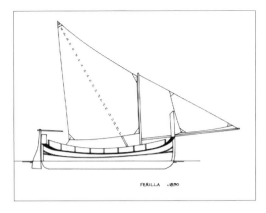

FERILLA -1890

then looked upon more and more as a plentiful source of supply to supplement these needs. The *ferilla*, in the eighteenth century, became larger and a rig and sails were fitted as they began to fish further out to sea. By the end of that century the *ferilla* was rigged with a spritsail. They were all double-enders – again from Sicilian influence – and were about 9 metres overall and carvel-built. The Maltese fishermen used the local measure of the *xiber* – equivalent to 10 5/16 inches (26 cm) – when referring to their vessels.

In the nineteenth century the *ferilla* developed into a popular fishing vessel and was to be found all around the island, and on Gozo. The spritsail became larger and a jib was rigged on a long bowsprit.

Shipbuilding was flourishing at this time under British rule, and this, in turn, helped the boatbuilding industry as training was given in the dockyard. Builders sprang up all around the coast of both islands, the highest concentration being around Grand Harbour itself. All the boats they built were of the same traditional construction and were similarly painted in bright colours, although different areas used different hues. They were all fitted with *falki* – removable washboards – which were removed when the nets were being set and hauled, and which were then slid back into place when sailing into the Mediterranean waves. The boats were carvel-built, frame first, with the centreline being set up to which frames fixed, these consisting of *majjiera* (floors) and *stamnair* (futtocks), before the planking was added. Traditional materials used in the construction were centuries old. Greenheart was imported from Britain for the keel, although this was often replaced by ash because of its high cost. Ash was also used extensively for the frames, with oak from Sicily occasionally being an alternative. This ash was imported from Yugoslavia and Albania. Planking was of red deal from Austria or Yugoslavia or pitch pine from America. Bulwarks were normally of mahogany and these were varnished, thus being the only part of the boat not painted. Today the same sorts of timber are still used, although white deal has replaced the red.

The rig was simple to operate. The sprit boom was housed into notches found on the thwart just abaft the fore *tambouret* or foredeck. The sail was hauled with a halyard through a block attached to the end of this boom. The small jib was fixed to the bowsprit, which was heaved into position, and the sail tightened through a sheave in the top of the short mast.

The *ferilla* was said to be renowned for its speed, being a fast and manoeuvrable boat. Like many fishermen throughout Europe, the Maltese held annual regattas in Grand Harbour and, prior to this, they would spend up to a week preparing their boats for the event. Competition was fierce, and the regattas were the highlight of the year, with all types of local boats being raced.

At the beginning of the twentieth century the small port of Marsaxlokk became the main fishing harbour of the islands and consequently the harbour was full of these *ferilli* (plural). Some were built there while others hailed from nearby Birzebbuga.

A painting of an early *ferilla* with rather romanticised *arznella* baskets, though it is plain to see that these are being set on a train of baskets.

A *speronara* from a painting by the same artist. The protrusion – the *sperone* – at the bow is clearly visible.

At the same time as the *ferilla* was evolving, so was the *Gozo boat*, a larger vessel. However, they had a forerunner, the *speronara* or *xprunara*, a double-ender which also originated in Sicily. In the sixteen century this vessel was referred to as the *brigantine of Gozo*, a boat that was both sailed and rowed. No details have surfaced as to its features before the eighteenth century, when it was rigged with one spritsail and a small jib. By the nineteenth century a bowsprit had been added and eventually the fishermen adopted the lateen rig on one, two or even three masts.

The *speronara* took its name from the *sperone* that was often found on Maltese boats, including the *ferilla*. This is a spur fixed onto the cutwater at the bow simply as a form of decoration with no practical use.

The *speronara* was also a fast vessel which ventured further out to sea than just Gozo. They commonly sailed over to Sicily or Italy, and to North Africa. From the early boats of around 15 tons, by the end of the nineteenth century there were boats with three masts of 50 tons. But around that time they drifted into obscurity due to developments with steam and other improvements such as the Gozo boat.

The *Gozo boat* appeared around 1880 and succeeded the *speronara* almost immediately. These were much deeper than the very shallow, flat-bottomed earlier boats, and were built up to about 40 tons. They were half-decked, and were built by the same technique as the *ferilli*, with similar bright paintwork and detachable washboards. By 1910 they had completely superseded the *speronara* and were to be found anywhere between Mjarr, Gozo's main harbour, and Grand Harbour. They survived long after the Second World War, only dying out after the modern vessels were introduced. The last boat that worked was, until recently anyway, sitting ashore at Mjarr. However, by that time they were only used for moving freight between the islands and it had been many years since one had fished.

Two smaller fishing craft operated around the islands. Both the *kajjik* and the *frejgatina* are representative of Malta's fishing craft. The *kajjik* originated from the *caiques* carried by the galleys of the Order of St John – between 1530 and 1802, the island was under the control of the Knights of that name. The modern *kajjik* is about 6–8 metres overall, transom-sterned and yet nowadays is a popular vessel. Brightly painted, as all the Maltese craft are, the transom stern was developed from the earlier pointed stern to facilitate the fishing with local traps and other methods of fishing. Some say this influence came from the British or, more likely, the Sicilians.

The *frejgatina* is the smallest of all the Maltese fishing boats. It is also transom-sterned, open and traditionally rowed with two oars. Nowadays some are engined, some even having a second engine in reserve or to drive the winch.

But it is the *luzzu* that has today become the symbol of Malta, alongside the Maltese Cross, and was even featured on the reverse of an older series of the country's lira coins.

The motor was introduced to the island's fleets around 1920. It was immediately found that the *ferilla* was unsuitable to receive these units because of its sharp form. Thus the *luzzu* appeared, once again introduced from Sicily (not Britain, as some have suggested in the past), modelled on the sprit-rigged *varca di conzu* which, as we saw in the last chapter, resembled the west coast *gozzi*. Whether it is coincidence that the name '*luzzu*' is a cross between '*luntro*' and '*conzu*' is unknown.

Almost immediately the flatter shape of the *luzzu* enabled engines to be fitted with ease. These vessels were altogether larger boats at 12 metres in length, much heavier and had a higher freeboard. Nevertheless, that there was some influence from the *ferilla* was obvious, given the *luzzu*'s appearance. Consequently, it was a slower boat. The first boats, although they were engined, retained the spritsail as the early engines were pretty unreliable, and they were also rowed by four oars when neither sailing nor motoring. At first the *luzzu* was a transport vessel rather than a fishing vessel, although within a few years the fishermen had latched onto and adopted the design. With their bright paintwork and stout shape, they were the perfect tool for the

A dramatic picture of a Gozo boat with two lateen sails on similarly sized masts.

A peaceful scene with a Gozo boat sailing along the coast. Note that all the Gozo boats have a 'G' on the sail.

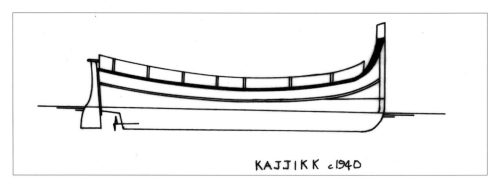

KAJJIKK c1940

A collection of small fishing craft – *frejgatina* and *kajjik*. The wire baskets are for catching octopus.

fishermen, who were reaching further and further afield in their search for fish. The bright colours were also believed to attract the fish.

Until the end of the twentieth century *luzzijiet* (pl) were built in various places around the islands: Kalkara (four builders), Marsaxlokk (two), Birzebugga (two) and Sliema (one) on Malta; and Mgarr (two), Victoria (one), Xlandi (one) and Marsalfarn (one) on Gozo. Today, only the Caruana brothers at Mgarr, with over 300 years of tradition behind them (their father used to work at Kalkara), are one of only a few left building wooden boats, though few of these new boats are *luzzijiet*. Today the *skuna* (schooner), again introduced from Sicily, is replacing many of the older fishing vessels. Several years ago the Caruana brothers were lengthening such a boat by 10 feet amidships, a common occurrence, by cutting the keel and scarphing in a new piece or replacing the entire keel, then fitting new frames and re-planking by cutting back every other timber.

Those *luzzijiet* that do survive now have up to four engines aboard. One central and two outer engines ensure total reliability while the fourth engine runs a small generator that in turn powers the winch and lights. Sweeps are retained, although these are more often than not used nowadays to carry a cover over the main part of the boat to afford some degree of shelter from the cold in winter. That being said, it appears that the majority of the boats are hauled out of the water in November or early December for the annual refit. Sitting on the shores, the fishermen scrape and sand them to bare wood, maybe re-caulk with *xoqqa* (oakum) if necessary, and paint them again, first using three or four coats of red lead before putting back their suit of bright colours. They are then put back in the water in the early part of the new year, ready to start fishing in March. Not many of the fleet fish in winter; these days

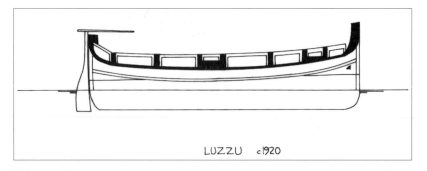

LUZZU   c1920

A group of fishermen standing proudly in their *luzzu*, possibly on its launching.

the fishermen either supplement their income by having another job or are involved in the fish farms that have recently appeared off the coasts. With declining catches, legislation from the European Union and escalating fuel costs, it's no wonder that the island's tourist industry has promoted the image of these boats. For it seems, not for the first time along Europe's coast, that taking parties of tourists out in the boats has proved more profitable than going fishing.

THE EYE OF OSIRIS

One feature on Maltese boats that must be mentioned is the 'eye' that is fixed either side of the high prow, sometimes delicately carved and carefully painted. This characteristic is found in many Mediterranean countries and is said to have been carried to Malta by the Phoenicians. It is believed that the eye wards off evil spirits and protects the boats, a custom going back many centuries. The eye is that of Osiris, one of the ancient Egyptian gods who was drowned and then torn into fourteen pieces before being scattered around the Earth by the god Seth. A new life was given to Osiris by the goddess Isis and from then on Osiris became the god of immortality and the underworld. The idea of the eye is that it enables the boat to see where it is going and hence avoid danger. Similar superstition forced fishermen to name their boats after Catholic saints, which they did until recently. Most still do although it is still possible to see boats with names like '*John F. Kennedy*' and '*Seabird*'!

FISHING TECHNIQUES

Some of the traditional Maltese fishing techniques are unique, other being typical of the Mediterranean. The main ways the fish are caught are:

1. Drift-netting for *Tumbrelli* – Frigate tuna or Frigate mackerel – which are caught between April and September with a drift-net up to 900 metres long.
2. One of the most unusual methods is used to catch *Lampuki* – the common dolphin fish – which is in season from September to December. The fish are caught using *kannizati* – floats with palm leaves attached. The palm leaves are cut and gathered from the large lower fronds of the palm trees, which they plant before the season. The *kannizati* are anchored in a particular place using stones. The *lampuki* shelter under the leaves and then are surrounded using a sort of seine-net that envelopes the catch. The fish themselves are only found in these waters, having hatched out in the waters of the Nile Delta and swarmed into Maltese waters in autumn.
3. Trammel-netting with fixed nets for bream and rock salmon. These nets are set on the seabed in the early evening, anchored and the top held up with floats. The fishermen then return in the morning to haul them in, hopefully clearing the catch that has become entangled in the layered net.
4. Large basket creels for *Arznella* – or picarel, a pelagic fish similar to pilchards. These are set in 20–30 fathoms of water, with herring inside as bait. The season is September to March.
5. Long-lining for *Pixxispad* – swordfish – with about 20 metres between hooks, and up to 1,000 hooks, i.e. over 10 miles long.
6. Wire traps for Octopus between March and November. The traps are either round or oblong, and are similar to shrimp cages.
7. Hand-lining for Tuna in winter, when the weather is fine, and there is not too much wind. As most of the larger boats are laid up, the *frejgatini* are used for this.

SOURCES

The Maritime Museum at Vittoriosa is well worth a visit and its curator Joseph Muscat has a wealth of knowledge on Malta's maritime past in general. Other than that, information came from talking to fishermen and the Caruana brothers.

CHAPTER 18

Croatia and the Former Yugoslavia Coast

The Gulf of Venice to the Ionian Sea

In Chapter 16 we saw how the *bragozzo* was the fishing boat of the northern part of the Italian coast, working from such ports as Trieste, Venice, Chiogga and Cesenatico, but they were also common in Istria (Slovenia) and along the Dalmatian coast of Croatia.

In Croatia, recently released from a horrific civil war, the fishing industry has grown over the last ten years and become a major source of revenue. While pollution from the River Po affects the Italian side of the Adriatic, and the centre is supposed to be a breeding ground left alone by fishers (although it often is not), the Croatian coast is largely pollution free. Most of the fish, ironically, is exported to Italy either by road or by being sold direct to Italian boats to end up in the Venice market as 'Italian fish'! Once Croatia joins the EU, its waters would be open to Italian boats, a move opposed by the fishermen for obvious reasons.

Today, compared with much of Europe, much less is known about ancient Croatian craft and nearly all of the original craft were destroyed by war and political dogma over the course of the twentieth century. Most coastal villages today have small fleets of open, motorised craft which the fishermen call the *batelo*, and larger fleets are based at Vis in the south, Zadar, the islands of Krk and Hvar (where at Sucraj the fishermen call their craft *kocha*) and, on the west side of the Istrian peninsula, at Pula, Rovinj and Porec, all of which were fishing harbours back in Roman times and probably before.

The traditional fishing vessel of the Istrian coast was once the lug-rigged *batana*, which ranged, on average, from about 4-6 metres

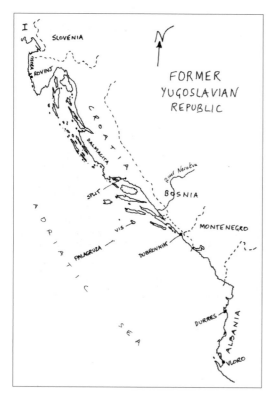

in length. These flat-bottomed craft, developed to fish in the shallow waters of that part of the coast around Rovinj, appear to have evolved from similar craft in the Venetian lagoon and evidence points to their use in northern Italian waters where a large version – the *batana grande* – had two lugsails much like the *bragozzo* of Italy and the largest of these can be up to 8 metres in length. There are records of these vessels working in the Italian Marché, including the lakes Trasimeno and Varano, and that these influenced the fishermen in the north of the Adriatic. Generally there are three types of *batana*, the *batana Risorta*, the fully decked boat; the *batana Classical*, the completely open smallest vessel; and the *semi-closed batana*, a half-decked version of which none exist today. The name '*batana*' comes, according to local tradition, from the Italian '*battere*' – to beat, hit or strike – as in the sound of the beating of the waves against the boat's flat bottom. There's a saying in Rovinj, today the home of the *batana* – '*Bone da bati mor!*', which translates to 'good enough to beat the sea!'. Some counter this and say that the name comes from the ancient maritime term '*batto*', which was used to denote a small boat with oars from the fourteenth century and which is the forerunner of the *batana*. That word, in turn, comes from the Old Saxon word '*bat*', the origin of the English word 'boat'.

Up until the 1920s the numbers of *batanama* (plural) working in Rovinj were relatively small and those that were there only fished locally. Once paraffin was obtainable for the lamps, and with the more frequent use of trawl nets for the night-time sardine fishing, there was a surge in numbers. From a traditionalist point of view, though not from that of the fisher, the golden age of this boat was the 1960s, when sails had disappeared, which in itself was brought about by the appearance of the Tomos 4 hp two-stroke outboard motor which was manufactured in nearby Kopar. Boats became versatile and faster than under sail. Few families didn't have a boat and those that did used them for leisure to get to a favourite swimming spot and do a spot of fishing. Most fished during the winter and before Easter they sought cuttlefish and squid for the traditional Easter meal. To date, 241 *batanama* have been listed in the Rovinj Port Authority Registers.

Above left: A *batana* from Rovinj, after 2004. (*Photo: Kuca o batini/Casa della batana*)

Above right: A *batana* in Rovinj in 2006, obviously recently restored. (*Photo: Robert Simper*)

Batanama have a slight rocker in the bottom and a slight convex shape in the stem. The framework is normally oak – Croatia did have plentiful supplies – while the planking was spruce or fir. The addition of a small transom was to accommodate the outboard motors. The lug – *rov vila al tierse* – was regarded as being large for the boat though they were as often as not rowed or, with a cut-out in the transom, sculled like the Venetian gondola. In the same way that the Italian fishers painted their sails, so did the fishers of Istria. The sails, of robust cotton, were painted in this way to ease the identification of a vessel and ninety-five families of Rovinj have so far been identified. In 1935 there were some fifty-two vessels with painted sails. By 1960 there were none. However, in 2004, the first new *batana*, 'Our Rigno', was launched, followed by the 16-foot *'Fiamita'* in 2005.

The fishermen of Rovinj are also renowned for their singing while mending their nets or fishing. These *bitanada*, traditional songs, were a way of imitating various musical instruments by voice so that, with many voices imitating various different instruments, they were able to perform as an orchestra at a time when, obviously, there were unable to use their hands. Instruments so copied were various guitars, the mandolin and mandola.

The *bracera* from the north of the country is another traditional type. This bluff-bowed, lateen-rigged vessel, it has been said, originated from the town of Brazza, which gave it is name. This vessel was common right along the coast, from the Gulf of Venice in the north and throughout both the Dalmatian and Istrian coasts. Upright in stem and stern, a typical *bracera* was 10 metres in length by 3.5 metres in the beam, half-decked with some decking over the after end of the vessel. They also set a foresail on a long bowsprit. One 10-metre replica, *'Our Lady of the Sea'*, was built by Mile Jadresic at his yard in Betine.

The *leut* – sometimes called a *trata* – was a widespread fishing boat throughout the Adriatic and was largely used in the pelagic fishery, whereby they dragged a sort of seine net across a bay. Rigged with one dipping lugsail, they were similar in length to the *bracer* and were renowned for their speed. One 1894-built *leut*, the *'Slobodna Dalmacija'*, was built in Korcula and was rescued and restored in 2008. The *leut*, it appears, had an exceptional seaworthiness under sail and an excellent rowing ability and was used for transporting goods as well the tool of some pirates working the coast.

Komiza, on the island of Vis, was the centre of fishing in Dalmatia from the sixteenth century up to the 1930s, even hosting an international fishing conference during the Austro-Hungarian era. After the collapse of the fishing, Vis became an Allied base during the Second World War, supporting the Yugoslav partisans under the leadership of Tito. He used a cave in the interior of the island as his campaign headquarters for several months. After the war a small naval base was installed there and the island was closed to all visitors until the 1990s. More importantly, though, is that Vis was home to a particular type of fishing vessel that has had a replica built. This is the *gajeta falkusa*, a vessel said to have originated from there. Vis is one of the nearest islands to Palagruza, a small island with a lighthouse situated out in the middle of the Adriatic and in the richest fishing area in the whole Adriatic, especially for sardines, although the fishermen also fished the nearer islands of Svetac and Jabuka. It's a 42-mile journey from Vis to Palagruza so the fishermen, venturing out when the

An old drawing of a *leut* landing a seine of fish.

A *leut* dated to 1925, possibly at Rovinj.

A lateen-rigged *leut* with the sail down off the southern part of the Croatian coast.

north-west '*maestral*' wind was blowing, had to take with them a full load of salt to cure their sardines once they were caught. Thus their boats had to be sailed as well as being capable of carrying a heavy cargo of fishing gear, salt and men. Usually they would sail but if the winds were adverse or light they would roll up the sail onto the yard, then they would row by standing upon boards – *balestrina* – laid across the boat and the five crew would all row. The *borka* was a smaller version of this boat which was rowed by three crew and seldom voyaged as far offshore as Palagruza.

Once at the island, when the moon was new, fishing could begin but they needed a light boat, easily manoeuvred with five oars, with a low freeboard to set and haul their gill-nets. Thus everything they didn't need for fishing was detachable. Once the fish was landed, and salted ashore, they had to wait for the '*jugo*' wind from the south to sail home fully laden. Because for the Komiza fishermen the winds were always favourable for their fishing expedition, and for other islanders it was not, given the geographical position of Vis, they came to command it and the thus the *gajeta falkusa* evolved over time to suit their purposes. To the normal light vessel are added *folke* or *falke*, detachable washboards that increase the freeboard. These and the mast and bowsprit are fixed using rope and dowels, and the cross-beams temporarily removed to allow salt to be packed beneath. Once they have arrived at Palagruza, the *folke*, mast, bowsprit, rig, rudder, barrels of salt, wine and water – everything that gets in the way of net handling – gets left ashore until the return journey and they use oars only while fishing.

There's a tradition that the fishermen, during their 20-day expeditions to Palagruza, used to climb the steep rocks of the island to collect firewood for cooking their sardines, while at the same time collecting the wild carnations for their sweethearts back home. This they kept secret for the older men forbade it, believing the ancient prophecy that the fishing would perish if the carnation disappeared from the rock of Jabuka. It did eventually and the fishing, which finished in 1936, was said to have lasted a thousand years. Most of the ancient *gajetas* were burnt during the festival of St Nicholas on 6 December each year as a tribute to the bounty from the sea.

Postcard of Komiza in the early twentieth century with a *gajeta falkusa*. (Photo: Velimir Salamon)

Fortunately, one *gajeta* had survived after being damaged in a storm in 1988 and beached on the island of Bisevo. This, the '*Cicibela*', was surveyed and the lines lifted, which enabled Komiza doctor Gordan Straka to build a scale model. Using paintings and photographs, the lines so lifted, and the scale model, as well as the fruits of 20 years' research, naval architect Velimir Salamon was able to produce a plan to enable the replica to be built in the shipyard at Trogir, on the mainland west of Split. At 9 metres in length, 2.9 in breadth and with a 9-metre-high mast, the vessel was built using pine from the island of Svetac, although the keel is oak. Authenticity was insisted upon at every stage so that the distilled pine resin (*katram*) was applied to protect the hull and spars from decay. The replica vessel was launched in 1997 and named '*Comeza-Lisboa*' for she was built in time to appear at the EXPO 98 in Lisbon. Two years later she appeared at the Brest maritime festival, where I had been lucky to have seen her. In 1999 she returned to Palagruza and fished in the age old way under oars, Today, though, I'm not sure exactly where she is.

Today, of course, most of the sardine fishing has gone, as have the seven canning factories on Vis, although there are newer ones on other Croatian islands. Tourism has replaced much of the former traditional way of life. Tuna is still landed, although the practise of closing a whole bay with a net, as once happened in natural bays such as the one near Kraljevica, to the north, is long gone. Sardines are still cooked for the increasing number of tourists that visit Vis but the daily tradition of grilling at exactly 1 p.m. each day, the air full of their aroma, has long disappeared. The 'Fishing Museum' in the sixteenth-century Venetian castle overlooking the south end of Komiza's harbour is worth a visit. Inside is the '*Cicibela*', lovingly restored, re-rigged and repainted by retired fisherman Ivan Vitaljic-Gusla, and, encouraged by Josko Bozanic-Pepe, it is now exhibited as a testament to these fine craft and those sailing and fishing them. However, hurry, because the charm and peace of this island will soon be swallowed up by tourism in the same way that the factory ships and hi-tech trawlers have destroyed the fishing industry.

Palagruza, by coincidence, is passed during the ferry crossing between Ancona and Patra, and only then can one gain the real sense of this out-of-the-way place. The

The *gajeta falkusa* built after taking the lines off the '*Cicibela*', here racing under full sail. (*Photo: Dinko Bozanic*)

lighthouse, built in 1875, is the biggest in the Adriatic and is rather a grand building, so much so that today there are now two four-bedroom apartments in the building that are rented out to holidaymakers keen to get away from it all. According to the website, all provisions must be bought before boarding the boat in Korcula, for you are on your own for the whole of the two weeks. The lighthouse is manned by a keeper so presumably there will be someone else to communicate with. The island is rocky with some degree of vegetation and home to many birds, and there are two fine pebbly beaches to swim off. And as you sip your evening wine, watching the sun set over the western horizon, there are two other factors to think about. Firstly, there were once some 500 fishing boats working the island, a not-insignificant number. Secondly, Pope Alexander III arrived here on Ash Wednesday, 9 March 1177, aboard one vessel in his fleet of ten galleys on their way to Venice via Vis. He had his dinner on the smaller island of Palagruza Mala (there's also another called Palagruza Velika), although why he chose to eat on the smaller island no one knows. Perhaps to be able to gaze back towards the main island and enjoy the views!

The *guc* or *gozzo* was the small, open double-ender with a short deck at both bow and poop. Used mostly in spear and line fishing, the boat was both light and fast under oar, though some set lugsails on two masts. However, it has been said that the term '*guc*', meaning 'swelling', refers to the fact that the hull is fuller than the Tyrrhenian Sea *gozzo*, with which there are many similarities. The *pasara* seems to be a transom-sterned *guc* and became popular as this allowed the fitting of an outboard motor after the 1920s.

Two river craft of importance were the *trupa* and *ladja* of the River Neretva, although the latter was simply a transport vessel or barge moving about the river delta, taking animals, fruit and vegetables from the fields, timber and the marriage dowry along or across the river, and are also used for funerals and other celebrations. Although smaller, the *trupa* was used for fishing, catching eels and frogs, leeches and bird hunting as well as some small-scale transporting. The *ladja* are up to 8 metres while the *trupa* is seldom larger than 4 metres. Both were either poled along the river or rowed while the loaded *ladja* was sometimes pulled alongside the riverbank by a group of people walking the bank. Both were capable of setting a small lugsail.

The *ladja* is probably the oldest of the type and evolved from the prehistoric dugout called *ladvas* which continued in use on the river up to the early twentieth century, the *serillias*, a stitched boat which has long disappeared from the river but which was in existence from the time of Christ, and the old Croatian *conduras* of the eleventh and twelfth centuries. Indeed, several archaeological ship finds off the port of Nin – just north of Zadar – have contributed to the understanding of these ancient boats which sailed all around the coast and out among the islands. They were stitched using flax rope and Spanish broom and caulked with organic material mixed with a tar substance. Carvel-built, they were up to 8 metres in length. Because leather has been found in proximity to the sunken vessels, it has been suggested that this was used for the sails. They had no keel and two longitudinal beams strengthened the hull, giving it a narrow, flat bottom which eased the beaching. The mast was set well forward, which suggested a small squaresail to catch the wind behind. They were also rowed.

Above left: Ladja lying in a shed. The shape of the hull is quite extraordinary.

Above right: A *ladja* (foreground) and a *trupa* (background) on the River Neretva.

The twentieth-century carvel-built *ladja* has always been a beamy boat at between 3.1 and 3.4 metres on an 8-metre boat (keel length: 3.4–4.1 metres) with a height of between 0.8 and 1.5 metres, which makes them look very flattish and unseaworthy. A bit like a floating oval dish! These boats, although displaying many similarities to the excavated *conduras*, especially by way of the framing, had a keel (*trup*) which was sunk into the soil on building so that 'it should not move'. The stem and sternpost were nailed to this and the ribs nailed either side. Five shortened ribs called *samotvorci* – literally self-shapers – would be nailed onto the stem and sternpost, these sometimes being referred to as 'little donkeys' in the Neretva valley. The top plank was fitted first to create some strength, and then the rest of the planking, working from the bottom upwards. A stringer – *stanza* – was bent round to strengthen the whole structure.

Although the dipping lugsail was efficient, the boats were normally pulled along the river, often by women, using a 50-metre-long rope. Occasionally the men did the pulling, and sometimes a horse. There were five oars, though only two were normally used, especially as the oars were the only means of steerage. These vessels – the real workhorses of the river – were covered in a black mixture of tar and pine wax, often proudly flying the Croatian flag. Some probably fished as well!

The *trupa* is a basic flat-bottomed canoe similar to many others of Europe (Flatner, Irish cot, Italian *sanaro* etc.). These simple boats have been described as the emblem of the River Neretva and few inhabitants along the bank did not have one. Trpimir Macan wrote of them: 'With the *trupa* the villagers used to row and sail, conquer the strong flow of the river Neretva, go through narrow canals and break through thick reeds. Sometimes the villagers used to carry the boat over the banks and say "sometimes I carry the boat, sometimes the boat carries me!"'

Built as many other flat-bottomed craft, the bottom is made up of either one or two planks, 24 mm thick, and to this the two-piece frames are nailed, five in total, as well as the stem and sternpost. The sides consist of one piece of timber 12 mm thick.

A *sandula*
on the beach
at Komiza.
(*Photo: Velimir
Salamon*)

The planking would either be pine, juniper or mulberry wood while the framing and stem and sternpost would be mulberry. The short mast was made from cypress and supported by a cross-member with its foot on a frame. Like the *ladja*, they were tarred regularly. Today they are constructed from plywood and painted. When there was no sail, a jacket or umbrella would be used to propel the boat along, though they were normally rowed or poled along.

Each year, on the second Sunday of August, the Neretva Boat Race is held when the *ladja* race a marathon from Metkovic (Croatia's Venice) to Ploce, a distance of 22 kilometres. This race is said to be the traditional competition of the descendants of the Neretvan pirates. However, today some 30,000 spectators turn out to watch the twenty-plus boats take part, each with twelve oars, the winner taking the 'Prince Domagoj Shield'.

The *sandula* of Komiza was a small, flat-bottomed boat used predominantly for inshore coastal fishing whereas the *gajeta falkusa, borka, guc, loja* and *leut* were open sea fishing craft. The *sandula* remained the typical Adriatic 'sharpy', not unlike the *batana*, which could be found all over the Mediterranean in various forms. It was built with transverse planking in the bottom, which became normal in Komiza during the years of isolation imposed by the Yugoslavian army (JNA).

SOURCES

The majority of this information comes from written evidence given to me by Velimir Salamon, naval architect and maritime historian, of Croatia. Some also comes from www.batana.org while John Robinson furnished me with some information on the *gajeta falkusa*. The Maritime Museum at Dubrovnik produced a booklet entitled *Experience of the Boat – Wooden Shipbuilding Heritage in Croatia*.

CHAPTER 19

Greece

The Ionian Sea to the Aegean Sea

The Greeks were among the first to develop the eastern Mediterranean techniques in boat construction. Ancient tradition suggests that the Cheops boat, discovered in Egypt in 1954, is 4,500 years old, though papyrus boats were in use much before that. Evidence from the Peloponnese refers to the discovery of obsidian in tool making, dated to 8000 BC, which had to be transported by sea from Melos. Homer, in the *Odyssey*, refers to the shell-first construction of vessels while the whole development of civilisation in this part of the Mediterranean is wrapped up in the maritime links that were made possible by their shipbuilding techniques. In terms of innovation, they were miles ahead of northern Europe. Shipbuilding really took off in the ninth and tenth centuries, during the Byzantine period, when the pioneering method of frame-first construction was developed. This way of building, as we've seen, spread quickly throughout the Mediterranean and eventually arrived in northern Europe a century or two later.

Archaeological evidence of the existence of eastern Mediterranean fishing comes from several excavated sites, though fishing scenes are much rarer in ancient Greece than in other cultures, a reflection of its lowly social status in society. In Egypt, net fishing is depicted on tomb reliefs dating from 2500 BC. These show seine-nets being hauled up on the beach after encircling a shoal. Presumably some sort of vessel was required to set the net offshore. Drying fish and the use of fish traps are also suggested. In

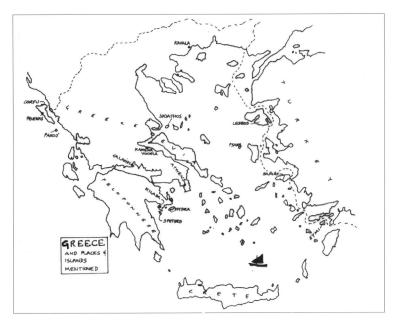

Greece the well-known fresco of a young fisherman of Thira dates from about 1500
BC while the Naxos pot discovered in a tomb from the Mycenaean era (1450–1100
BC) shows men hauling a seine-net with fish contained inside. At the archaeological
site of Kynos, on the North Euboea Gulf on Greece's eastern coast, a similar pot has
recently been discovered. Another pot depicts one of the first fishing boats in use on
this coast, though the earliest depiction of such a vessel comes from Crete and is dated
from the twelfth century BC. Several small clay models have also been excavated and
all this evidence has been judged as representing small fishing craft, probably crewed
by one man and used to fish close to the shore. Oppian of Coryrus, the Greek author,
wrote on sea fishing, producing the poem *Halieulica* between 177 and 180, and this
work survives today. In it he mentions that the fishermen used 'very light nets of
buoyant flax' used for tuna ... they 'wheel round in a circle round about while they
violently strike the surface of the sea with their oars and make a din with the sweeping
blow of poles' ... the fish are thus frightened and chased into the net. This confirms
that they were working nets from boats then. Also, a watchtower was manned to
search out and point the boats in the direction of the shoal, a practice still common
among the Cornish and Spanish fishers of the twentieth century. A third-century
African mosaic shows fishing boats with large square-sails. Another fourth-century
mosaic from Carthage depicts two persons fishing from a boat with a mast and two
stays. At the same time, the fishermen of Athens were said to work at night because
the catch kept longer, which allowed them time to supply the market at Athens in the
morning. Generally, the first boats worked in the channels through which fish swam.

As the fortunes of these fishers improved, man gradually built larger craft and
ventured further away from the safety of the shore. Thus, with Greece being
geographically a mountainous land where the vast majority of the population
inhabit the coasts or the immediate hinterland (half the population live in Athens
alone), fishing became both a major contributor to the economy and a principle
pastime to supplement people's self-sufficiency. Fishing boats, as elsewhere, developed
through improvements in fishing technology. Furthermore, with it having one of the
longest coastlines in Europe, partly due to having 2,500 islands, and a relatively low
population (12 million), it must surely have one of the highest density of fishing boats
per head of population.

BOAT TYPES

Today the *trechandiri* is the ubiquitous fishing boat of Greece. These are double-ended
boats, curved in the sternpost and having a straight raked stem, and are renowned
as being extremely seaworthy. Sizes vary, depending on usage, fishing method and
the nature of the shore from which they work, but generally range from about 6
metres to 10 metres. Larger *trechandiria* (plural) – up to 40 tons – were used for
trading. Trechandiria traditionally set lateen sails, though a change occurred sometime
in the nineteenth century when the larger versions adopted the *psatha* or lugsails

TRECHANDIRI

on two masts while the smaller fishing trechandiria adopted the *sakkoleva* or type of spritsail. The *psatha* could drive a vessel closer to the wind while the *sakkoleva* was deemed more suitable for rough waters. The majority of both versions of these boats were, and still are, decked over though there's little space under the deck in the fishing craft, for they have a relatively shallow draught for working close inshore. On top of that they are fast vessels and have a good carrying capacity, perfect for fishing.

There are two schools of thought on their origins. Some say they developed from a particular type of *caique*, first built in Hydra in 1658, while others suggest that they evolved from the *trabaccolo*, a type of sailing vessel used for trading in the Adriatic. A *caique*, though, is not a specific type of Greek fishing boat. It is a generic term for a vessel used for trading, and can also be used to include professional fishing craft. *Caiques*, then, can range anywhere between 15 and 200 tons.

Motors were adopted in the 1920s and within a decade or two the rigs had disappeared. The hulls became fatter, with a fuller body section, though the profile remained the same. Once the rig had been removed superstructures appeared and many of the smaller boats have cumbersome and ugly structures perched on their decks. Shelter from the hot sun, though, is essential and many have metal frameworks covered in canvas that adorn the rear end. Today – as is attested by the numerous postcards on show around the tourist areas – trechandiria are still very popular. Up until about 1970 they were being built in almost any coastal community. Many were built under the shade of a suitable tree with only hand tools being used, and others in gardens. Boatyards still decorated many harbours. Spetses is a good example, where six yards around the harbour were still building and repairing wooden craft recently.

In the ancient tradition, they were and still are constructed frame-first from pine. The best pine available – Aleppo pine, which is also known as Mediterranean pine (*Pinus brutia*) – was brought in from the island of Samos though, after wide-ranging fires on the island, Samos pine is today in very short supply. Spetses also had its own supply of this pine and, attracted by these trees, boats have been built here for centuries. Sadly, the forests also suffered from devastating fires in the 1990s. Another common factor of today's trechandiria which the postcards focus on is their characteristic colours, white hulls, bright superstructures, yellow fishing nets with contrasting red floats. Today, they are easily recognisable ambassadors for Greece!

All trechandiria have one thing in common, a length/beam/depth ratio of round about 9:3:1. During their building, once a backbone of the desired length is set up using these proportions, the process of creating the shape of the vessel in the frame-first construction begins. There are three ways to do this.

An unusual form of the sprit rig on a *trechandiri* hull at Pasalimani (now in Turkey) in 1907.

A boatbuilder at work on a small *trechandiri* at Patras in 2005. These craft are very still much in use.

The first and the oldest was developed in Greece and is called the *monochraro* or mathematical way. The shape of each frame is determined by mathematically decreasing the shape by increments from the designed midship section. This method was established in ancient times from Euclid's geometrical rules. In later times, this method developed into what is now called the whole-moulding method, in which three aids are needed – the rising square, the breadth mould and the hollow mould. Each had sirmarks for each frame and these are aligned to draw out the shape on the mould loft floor or directly onto the timber.

The second method is perhaps the most common, especially in larger shipyards and outside Greece, whereby the shape is scaled up from drawings directly onto the mould loft floor before being picked up and transferred onto timber. Half models can be used in this case as well. In the third way two or three frames are set up, their shape being pre-determined, and ribbands (longitudinal strips of timber) are fixed between these and either end of the vessel, and frames fitted in between, up to the ribbands.

Once the frame is completed, planking can begin. At the Lekkas Yard in Kilada, in the south of the Peloponnese, where they were building 25-metre vessels in 2001, it was a surprise to find that the planks were only some 30 mm thick, and that these were nailed using galvanised spikes. At the time they were building three vessels of different sizes, while across the road the yard of Basimakopoulis & Son was building a *Liberty*-type vessel, a cross between a *karavoskaro* – characterised by its clipper bow and used as a trading vessel – and the American liberty ships introduced into Greece after the Second World War. Basically, these Liberty-types have the clipper bow and a cruiser stern which was considered good for trawling. In between, these craft look very similar to the trechandiria and were first introduced into the offshore fishing fleets in the 1960s.

Another type of fishing boat is the *trata*, a boat similar in shape but with a wider beam. At one time the *trata* was the most popular oar-driven fishing boat and its name comes from the net the fishermen drag out from the shore to encircle a shoal. They were usually 8–12 metres long, although the largest were 15 metres.

A *tserniki* or *saita* is a vessel similar to a *trechandiri* but with a straight but raking stem. Only a very few remain in Greece, the oldest, at 90 years, still in Hydra. The owner explained that the name *saita* means 'arrow' and is so-called because the boat supposedly resembles an arrow.

Hydra also has its own *gaita*, a small, open boat used for fishing. These are unusual in that they have a plumb upright stem and a transom stern, and are about 4–5 metres long. In other parts of Greece, a *gaita* is simply a double-ended local boat with a flat bottom. *Varkalas* were larger vessels with transom sterns, often with decorative carvings, raking stems and often a schooner rig. *Barque*, on the other hand, is a generic term referring to any number of miscellaneous fishing vessels with sloping stems. A *bratsera*, I was told in Hydra, too, was unique to that island, but I've since found out that the term refers specifically to a vessel, more often than not a trechandiri hull, setting a psatha rig!

The *karavoskaro*, like the *varkalas*, was often schooner-rigged. The hull was, as mentioned, unique in that it had a clipper bow and a counter stern. The *perama* (*peramata* in the plural) are similar to the *trechandiria* in hull form except for the raking stem, which is straight or with the smallest amount of curvature. It also has a small vertical transom, or breakwater bulkhead, just abaft of the stem, to which the fore ends of the bulwarks are fastened. They are mostly used for trade. A few examples still exist, such as 'Elleni', built by Giorgos Mytilineos at Skiathos (this yard was near the present day harbour), which has recently undergone a refit. Two other *peramata*, 'Evangelistria' and 'Faneromeni', the latter also built in Skiathos in 1945, have recently been restored to their former glory.

Another type, the *gatzao*, is a sort of hybrid of the *trechandiri* and *trabaccolo*, the latter being a larger lug-rigged vessel with a high and bluff bow from the Adriatic, and was also in use among the Ionian islands and parts of the west coast of Greece. They were similar in size to a typical *trechandiri* although broader in beam.

Several types of flat-bottomed boats can also be found in Greece. On the west side, the Messolonghi lagoon is a shallow area with good fishing. Indeed, it was first inhabited

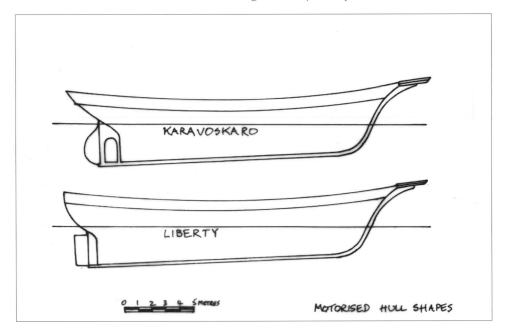

Above left: Building a Liberty-type boat at Kilada in the south of the Peloponnese in 2001. Today, though, that yard is devoid of new builds.

Above right: A *gaita* from Hydra, rigged with a lateen sail.

by Dalmatian pirates, who found a safe haven with excellent fishing potential. They built reed huts on stilts on the lake until they moved ashore. To fish the fishermen adopted a shallow-draughted canoe and today many of these exist on the waterways, fishing the many fish weirs on the shallow water. Most appear to be of up to some 9 metres in length, very narrow on the beam but with a high sheer at the bow. The stem itself is raked and either straight or very slightly concave while the sternpost is straight and raked, though less than the stem. Traditionally these were rowed or poled along, though today many now have inboard engines in a central housing and some are now used for pleasure purposes as against commercial fishing. There are distinct similarities between these canoes and the *batana* of north Croatia and one obviously wonders whether the Dalmatian pirates had any influence in the development of these craft. Meanwhile, there are still various small houses built on stilts out in the lagoon.

The *kourita* is another flat-bottomed craft that originated from Asia, where they were popular around Smyrna and the surrounding lagoons. When the Greeks were forced out of the area by the Turks in the 1920s, some of the vessels were brought to Greece. Many examples can still be seen on the island of Chios, while several have recently been seen at the small fishing harbour of Livantes on the North Eubeoa Gulf (close to the aforementioned archaeological site of Kynos), where one was measured by the author. At 5.4 metres long and 1.5 metres in breadth, the boat had been built in Chalkida in approximately 1960. Today she's engined with a Lister 7 hp unit which the current owner – who had inherited the boat from his father – thought had been installed at building. Flat-bottomed craft also operated in several of the Greek inland lakes.

The sponge-fishing vessels of the Aegean, the earliest recorded example coming from the island of Symi in the sixteenth century, are among the most spectacular of the Greek traditional fishing craft. However, there are two types of sponge-fishing craft – those that are used to dive for the sponges and those that use mechanical means.

The oldest way of collecting sponges, before the advent of diving bells and breathing apparatus, was by free diving, using a stone as a weight. Thus evolved the *skafi* or *skaphi* of Symi. Rigged with a large *sakkoleva* and foresails, these craft were fast and seaworthy. They had to be, for some sailed over to Egypt and others, it has been suggested, went as far as Tunisia in search of sponges. Some of the craft even had a tiny mizzen sail to aid with the steering. They had a transom which raked sharply, an even more pronounced rake in the stem to give a short keel length, a degree of sheer in the deck-line and a very deep draught. However, once diving equipment appeared, their use was phased out. At first, diving pumps were used and *trechandiria* were adapted, becoming known as *achtarmas* or *michanokaiko* (literally mechanical *caiques*). One of these latter craft, the 'Konstantinos & Eleni', built in 1936 and one of the last to fish for sponges, was decommissioned in 1994 and cut up as part of the enforced scrapping. Like in Britain, indeed throughout Europe, this disastrous policy has led to the destruction of much of Greek fishing heritage.

Later vessels that used trawls to collect sponges were called *gagava* and again were variants of the *trechandiri*, *varkalas* or *karavoskaro*, the latter being a trading vessel with a clipper bow with a pronounced protrusion, like a short bowsprit, and a counter stern. This method of trawling – also called *gagava* – involved dragging an implement

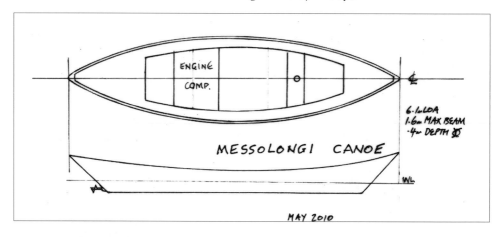

ENGINE COMP.

MESSOLONGI CANOE

6.1m LOA
1.6m MAX BEAM
.4m DEPTH ⌀

WL

MAY 2010

Above: Fishing at Messolongi with a drop net.

Left: Messolongi canoes still work the shallow water in 2011. Here, the fisherman is repairing his gas lantern.

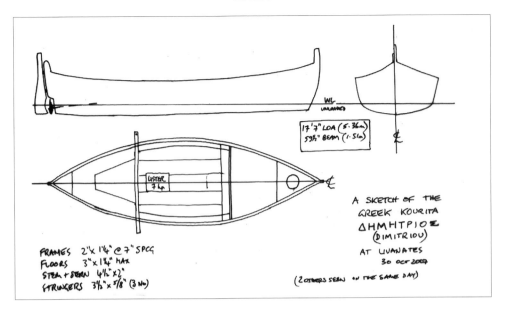

FRAMES 2"x 1¼" @ 7" SPCG
FLOORS 3"x 1¼" MAX
STEM + STERN 4½"x 2"
STRINGERS 3½"x ⅝" (3 No)

WL
UNLOADED

17'7" LOA (5.36m)
59½" BEAM (1.51m)

A SKETCH OF THE
GREEK KOURITA
ΔΗΜΗΤΡΙΟΣ
(DIMITRIOU)

AT LIVANATES
30 OCT 2007

(2 OTHERS SEEN ON THE SAME DAY)

Using a small *kourita* to service the fish weir at Aghios Konstantinos in the 1950s. (*Photo: Petros Kounouklas*)

A *trechandiri* adapted as a sponge boat in the 1950s.

Kalymnos, the home of the sponge boat, in the 1950s with an assortment of craft including, in the main, *peramata* and *trechandiria*.

along the seabed that uprooted and collected the sponges. In recent times this has been restricted because of the damage it does to the natural habitat of the seabed. Others, using harpoons, had transom-sterned *giala* which were much smaller at up to 7 metres long. The folk of Hydra say that their *gaita* developed from these small *giala*.

Once motorisation had fully been accepted (steam never had much effect in the Mediterranean) and had spread among the islands, larger *michanotrates* or motor trawlers evolved to trawl their nets called *gripos* with their large motors. Thereafter, *trata* generally referred to the nets set from beaches whereas a net dragged by two sailing boats was referred to as *anemotrata*.

SOURCES

There's a new fishing museum at Perama, though the exhibits were limited during our visit. There's also the Hellenic Maritime Museum in Piraeus, though it hasn't been open during our visits. Information came largely from personal communication and the books by Kostas Damianidis, who was instrumental in setting up the Perama museum. One of these – *Maritime Tradition in the Aegean* – has an English translation. Greece abounds with small museums, many of which touch on fishing, and the museum in Samos is worth visiting. Boatyards are still working and a visit to one of these is still invigorating and atmospheric, probably more so than any others in coastal Europe. Traditional *trechandiria* still fill most harbours, though plastic boats are quickly surplanting them.

Index